Collins

KS3 Maths Year 9

Leisa Bovey, Am
Katherine Pate

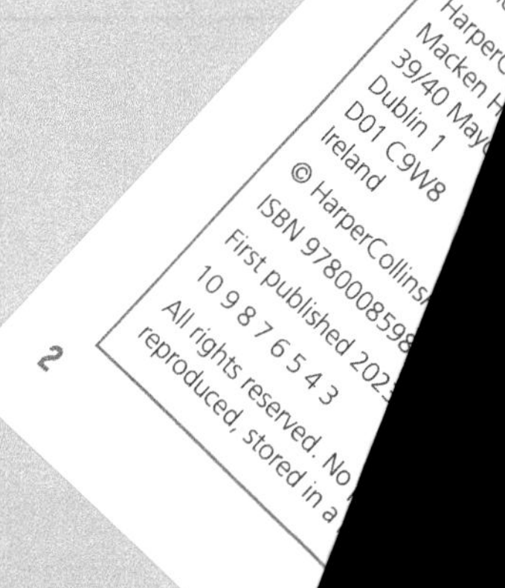

How to use this book

Each Year 9 topic is presented on a two-page spread

Organise your knowledge with concise explanations and examples

Key points highlight fundamental ideas

Notes help to explain the mathematical steps

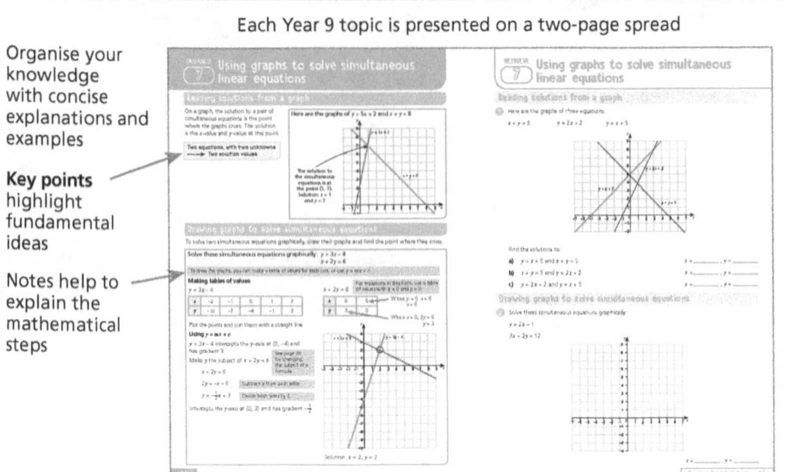

Test your retrieval skills by trying the accompanying questions for the topic

Mixed questions further test retrieval skills after all topics have been covered

Key facts and vocabulary section helps to consolidate knowledge of mathematical terms and concepts

Answers are provided to all questions at the back of the book

ACKNOWLEDGEMENTS

The authors and publisher are grateful to the copyright holders for permission to use quoted materials and images.

Every effort has been made to trace copyright holders and obtain their permission for the use of copyright material. The authors and publisher will gladly receive information enabling them to rectify any error or omission in subsequent editions. All facts are correct at ime of going to press.

l images ©Shutterstock and HarperCollins*Publishers*

lished by Collins
nprint of HarperCollins*Publishers* Limited
lon Bridge Street
SE1 9GF

llins*Publishers*
ouse
r Street Upper

662

British Library Cataloguing in Publication Data.

A CIP record of this book is available from the British Library.

Authors: Leisa Bovey, Ama Dickson and Katherine Pate
Publishers: Clare Souza and Katie Sergeant
Commissioning and Project Management: Richard Toms
Editorial contributor: Amanda Dickson
Inside Concept Design and Layout: Ian Wrigley and Nicola Lancashire
Cover Design: Sarah Duxbury
Production: Emma Wood
Printed in the United Kingdom

Contents

1 Similarity

Calculating the scale factor

Similar shapes are **enlargements** of one another.
Shapes that are similar have these properties:

- Their angles are equal.
- They have the same shape but are different sizes.
- Each side length has been multiplied or divided by the same number (called a **scale factor**).

The scale factor is the multiplier that is used to enlarge a shape. Each side of the shape is enlarged by the same scale factor to create a similar shape.

a) Shapes A and B are mathematically similar. What is the scale factor of the enlargement from A to B?

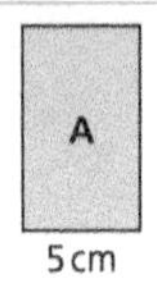

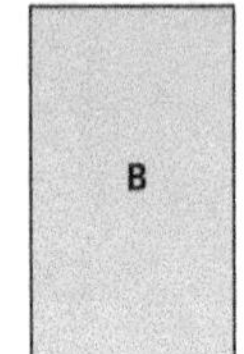

To find the scale factor, work out what number 5 is multiplied by to result in 20.

Algebraically:
$5 \times x = 20$
$x = 4$
The scale factor is 4.

b) Shapes X and Y are mathematically similar. What is the scale factor of the enlargement from X to Y?

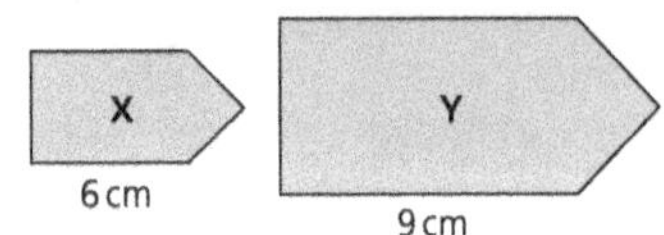

$6 \times x = 9$
$x = \frac{9}{6} = \frac{3}{2}$
The scale factor is $\frac{3}{2}$

An enlargement does not always mean that the shape gets bigger. If the scale factor of enlargement is less than 1, the new shape is smaller than the original. The two shapes are still similar.

Shapes M and N are mathematically similar. What is the scale factor of the enlargement from M to N?

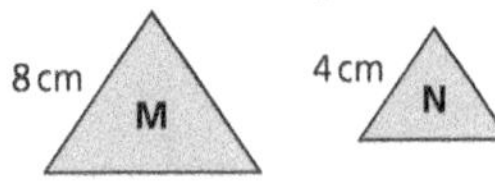

$8 \times x = 4$
$x = \frac{4}{8} = \frac{1}{2}$
The scale factor is $\frac{1}{2}$

Finding the missing length

To work out the missing length of a shape that is similar to another and is larger, **multiply** the scale factor by the corresponding length of the original shape.

To work out the missing length of a shape that is similar to another and is smaller, **divide** the original length of the shape by the scale factor.

Shapes A and B are mathematically similar. Calculate the unknown length in shape B.

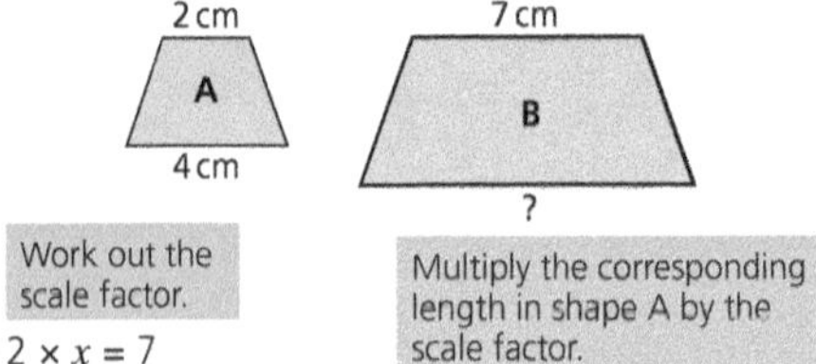
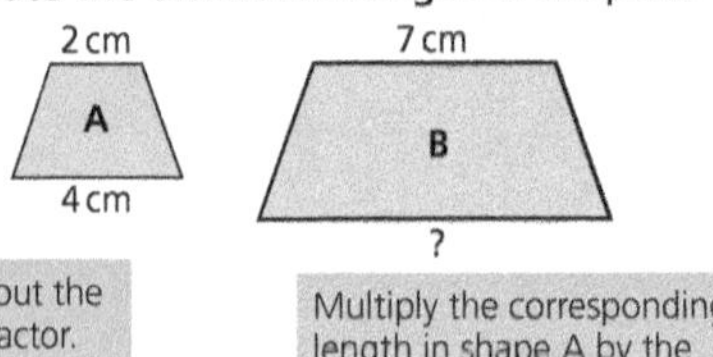

Work out the scale factor.

$2 \times x = 7$
$x = \frac{7}{2}$
The scale factor is $\frac{7}{2}$

Multiply the corresponding length in shape A by the scale factor.

$4 \times \frac{7}{2} = 14$
Unknown length = 14 cm

Shapes C and D are mathematically similar. Calculate the unknown length in shape C.

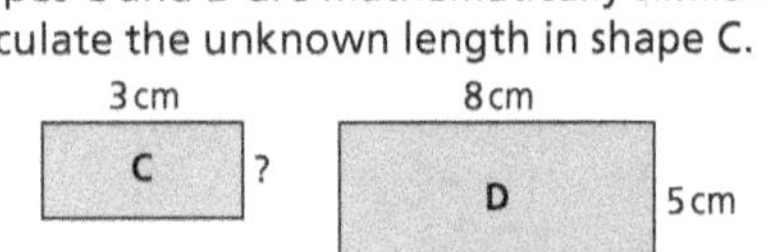

Work out the scale factor.

$3 \times x = 8$
$x = \frac{8}{3}$
The scale factor is $\frac{8}{3}$

Divide the original length by the scale factor.

$5 \div \frac{8}{3} = \frac{15}{8}$
Unknown length = 1.875 cm

RETRIEVE

1 Similarity

Calculating the scale factor

1. Shapes A and B are mathematically similar.

   Work out the scale factor of the enlargement from shape A to shape B.

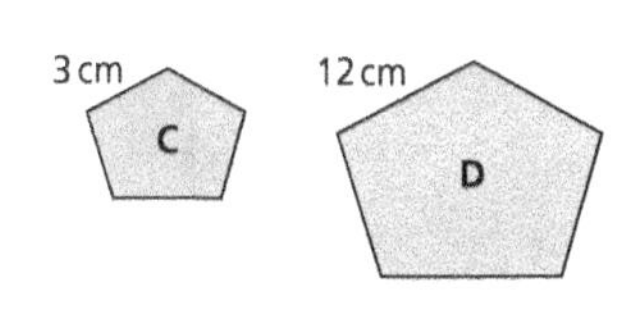

2. Shapes C and D are mathematically similar.

   Work out the scale factor of the enlargement from shape C to shape D.

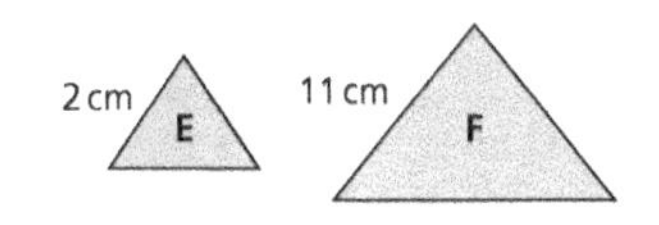

3. Shapes E and F are mathematically similar.

   Work out the scale factor of the enlargement from shape E to shape F.

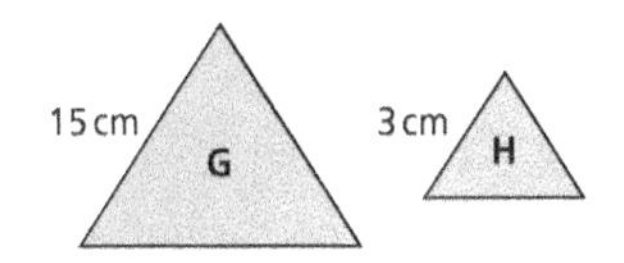

4. Shapes G and H are mathematically similar.

   Work out the scale factor of the enlargement from shape G to shape H.

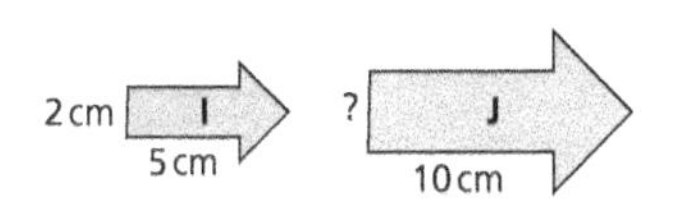

Finding the missing length

5. Shapes I and J are mathematically similar.

   Work out the unknown length in shape J.

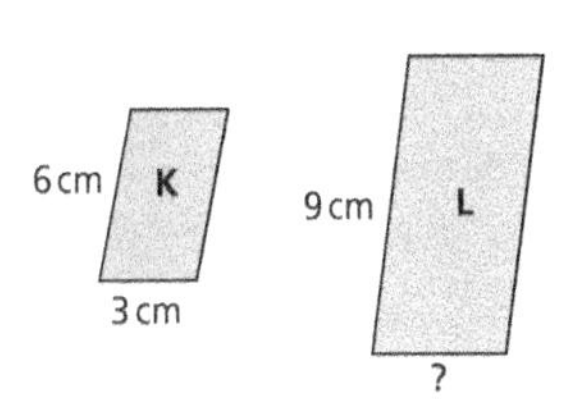

6. Shapes K and L are mathematically similar.

   Work out the unknown length in shape L.

7. Shapes M and N are mathematically similar.

   Work out the unknown length in shape M.

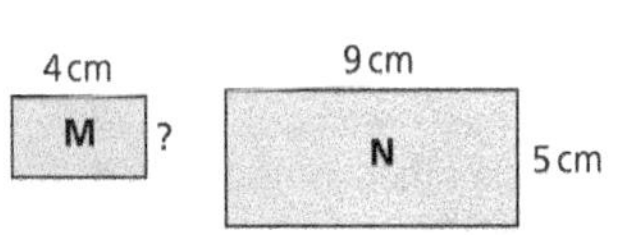

1 Congruence

Congruent shapes

Congruent shapes are **identical**. They are exactly the same size and exactly the same shape.

If a shape is reflected, rotated or translated, the image is congruent to the original shape. The corresponding sides and the corresponding angles stay the same.

Which pairs of shapes are congruent?

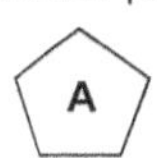

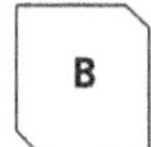

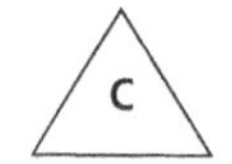

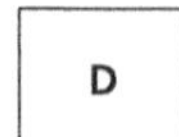

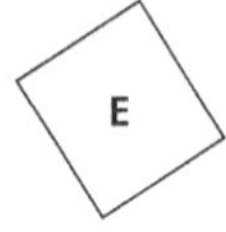

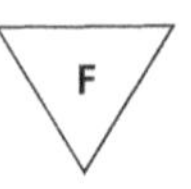

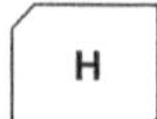

The following pairs of shapes are congruent:

A and G B and H C and F D and E

In each pair, the shapes are identical in size and identical in shape.

Congruent triangles

Two triangles are congruent if one or more of these four criteria are true.

Side, Side, Side (SSS)	Side, Angle, Side (SAS)	Angle, Side, Angle (ASA)	Right angle, Hypotenuse, Side (RHS)
The three sides are equal.	Two sides are equal and the angle between them is equal.	Two angles are the same size and a corresponding side is the same length.	A right angle, the hypotenuse and a corresponding side are equal.

Decide if the triangles in each pair are congruent, giving a reason in each case.

a)

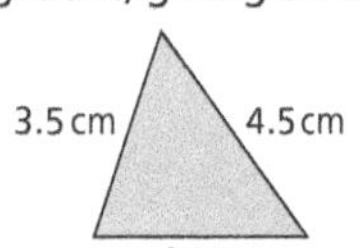

Yes, the triangles are congruent. All three sides are equal (SSS).

b)

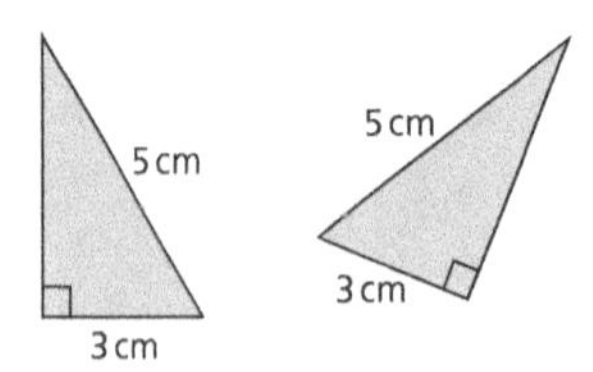

Yes, the triangles are congruent. They each have a right angle, they have equal hypotenuses and one other corresponding side is also equal (RHS).

c)

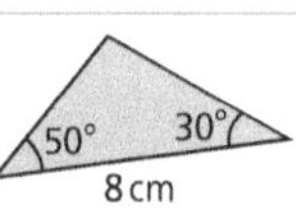

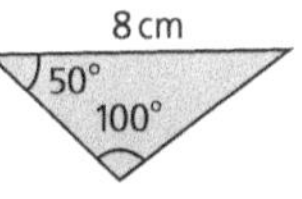

Yes, the triangles are congruent.

Angles in a triangle sum to 180°.

In the first triangle, $50° + 30° = 80°$ and $180° - 80° = 100°$ so the unmarked angle is 100°.

In the second triangle, $100° + 50° = 150°$ and $180° - 150° = 30°$ so the unmarked angle is 30°.

Therefore, the corresponding angles are equal. A corresponding side (the side opposite the 100° angle in each triangle) is also equal, so ASA is true.

d)

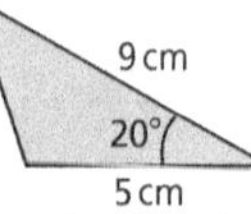

Yes, the triangles are congruent. They have two equal sides and the angle between them is equal (SAS).

RETRIEVE

1 Congruence

Congruent shapes

1. Find four pairs of congruent shapes.

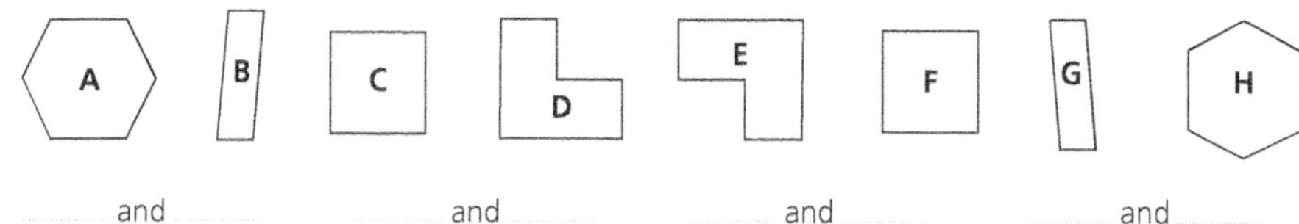

.... and .... and .... and .... and ....

2. Are these two shapes congruent?

Give a reason to support your answer.

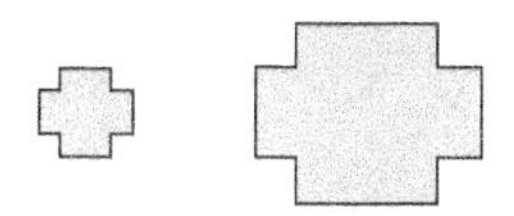

....

Congruent triangles

3. Are these two triangles congruent?

Give a reason to support your answer.

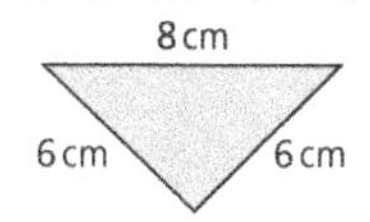

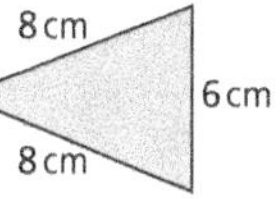

....

....

4. Are these two triangles congruent?

Give a reason to support your answer.

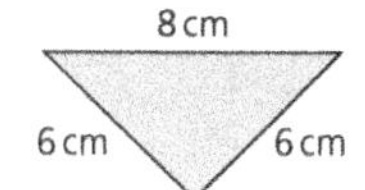

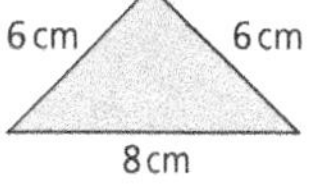

....

....

5. Are these two triangles congruent?

Give a reason to support your answer.

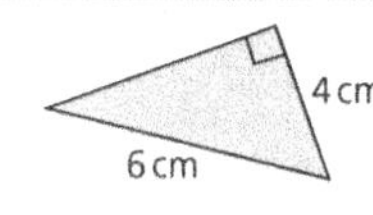

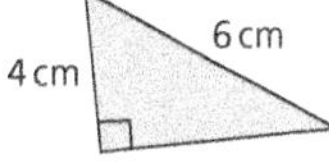

....

....

6. Are these two triangles congruent?

Give a reason to support your answer.

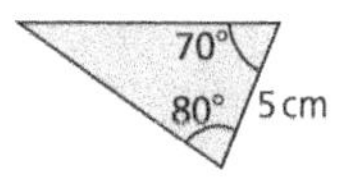

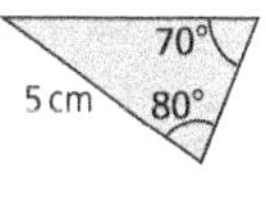

....

....

7. Are these two triangles congruent?

Give a reason to support your answer.

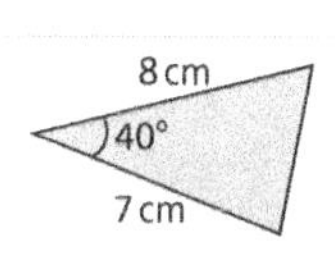

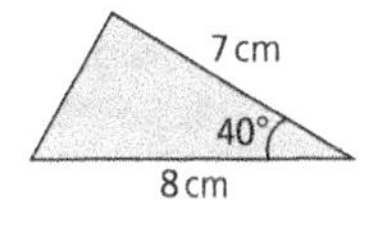

....

....

1 Symmetry

Lines of symmetry

A 2D shape is **symmetrical** if a line can be drawn through it and the shape looks exactly the same on either side of the line. A line drawn through a shape to show that it is symmetrical is called a **line of symmetry**.

A line of symmetry can also be called a **mirror line** because, when a mirror is placed along it, the reflection looks exactly the same as the original.

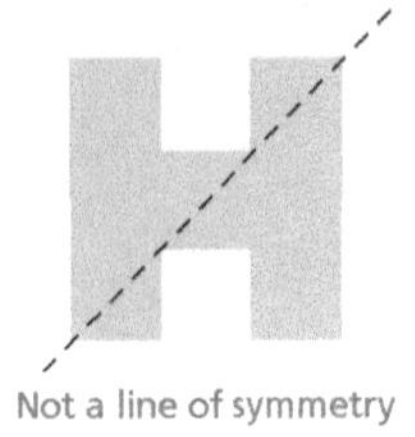

Not a line of symmetry

A line of symmetry

The number of lines of symmetry in a regular polygon is always the same as the number of sides.

An equilateral triangle has three lines of symmetry.	A square has four lines of symmetry.	A circle has an infinite number of lines of symmetry.
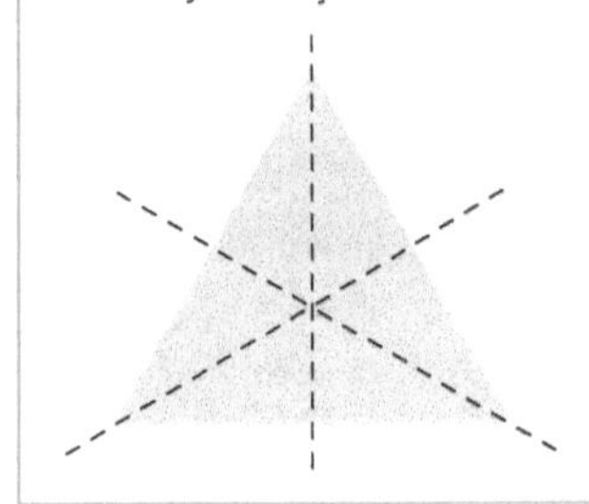	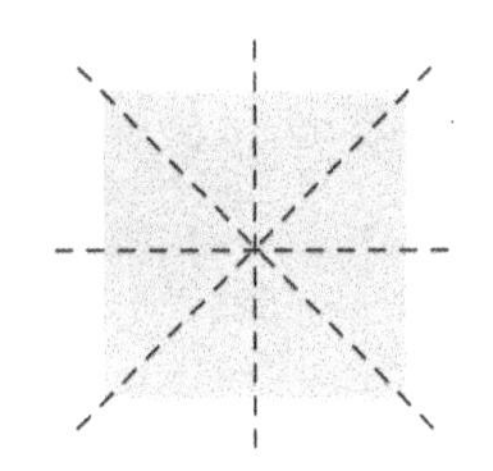	

Rotational symmetry

Rotational symmetry is the number of times a shape can be rotated about its centre for 360° and look exactly the same. The number of times the shape is rotated is called the **order**.

If a shape only fits into itself once, it has no rotational symmetry. This means that it has rotational symmetry of order 1.

A shape with rotational symmetry of order 1

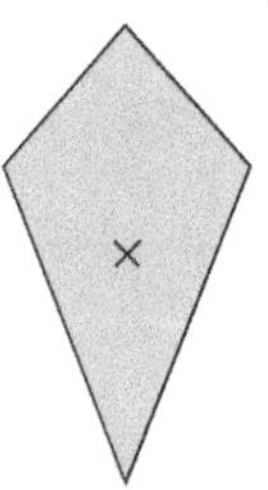

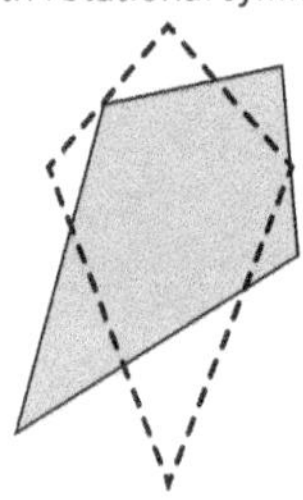

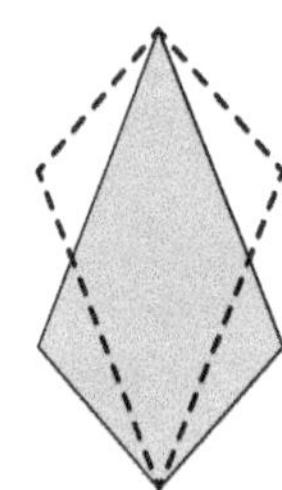

What is the order of rotational symmetry of an equilateral triangle?

An equilateral triangle can be rotated and fits into itself three times before returning to its original position.

An equilateral triangle has rotational symmetry of order 3.

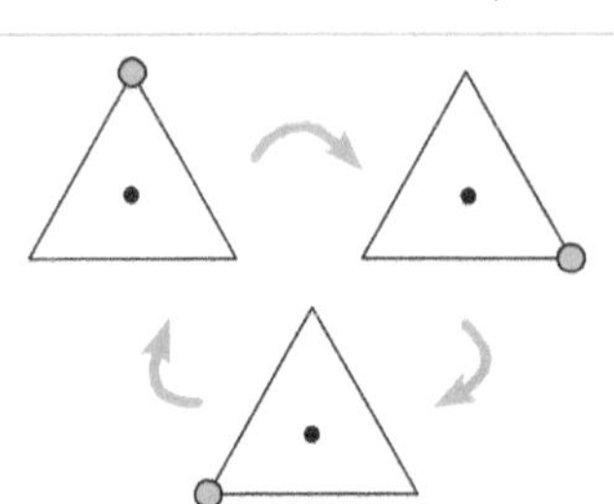

RETRIEVE

1 Symmetry

Lines of symmetry

1. Draw the lines of symmetry of a kite and state how many there are.

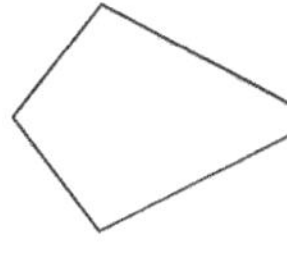

.... line(s) of symmetry

2. Draw the lines of symmetry of a regular pentagon and state how many there are.

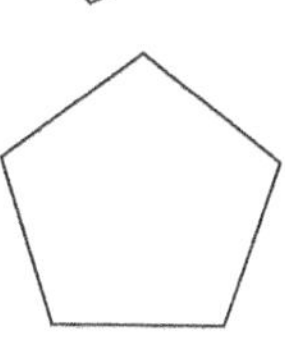

.... line(s) of symmetry

3. Draw the lines of symmetry of an isosceles triangle and state how many there are.

.... line(s) of symmetry

4. Draw the lines of symmetry of the shape and state how many there are.

.... line(s) of symmetry

Rotational symmetry

5. What is the order of rotational symmetry of a rectangle?

6. What is the order of rotational symmetry of a regular hexagon?

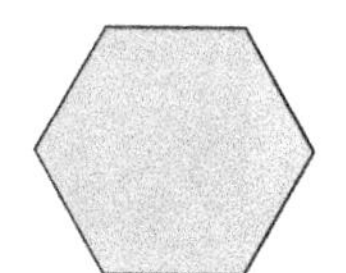

7. What is the order of rotational symmetry of the shape?

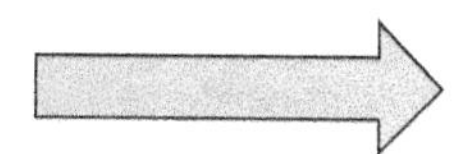

Pythagoras' theorem 1

Finding the hypotenuse

The hypotenuse of a right-angled triangle is:

- the longest side
- opposite the right angle.

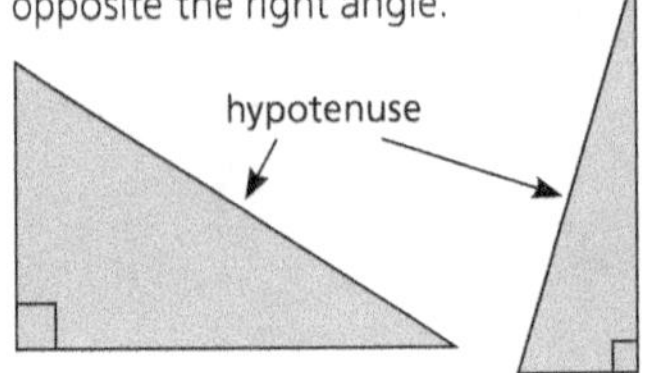

Pythagoras' theorem

$c^2 = a^2 + b^2$

c is the length of the hypotenuse

b

a

a and b are the lengths of the two shorter sides

Below is a right-angled triangle ABC.

Find the length of AC.

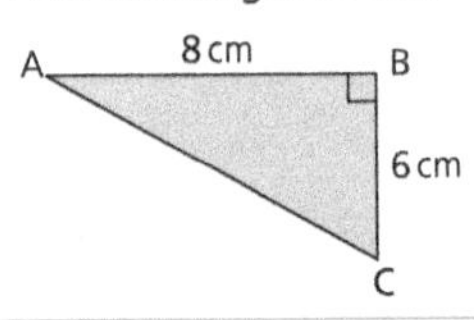

Label the hypotenuse c and the other two sides a and b.

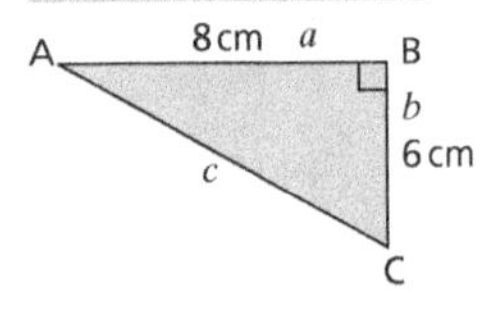

$c^2 = a^2 + b^2$ Write Pythagoras' theorem.

$c^2 = 8^2 + 6^2$ Substitute the values for a and b.

$c^2 = 64 + 36$

$c^2 = 100$

$c = \text{AC} = \sqrt{100} = 10\,\text{cm}$ Square root both sides.

Below is a right-angled triangle DEF.

Find the length of EF to 1 decimal place.

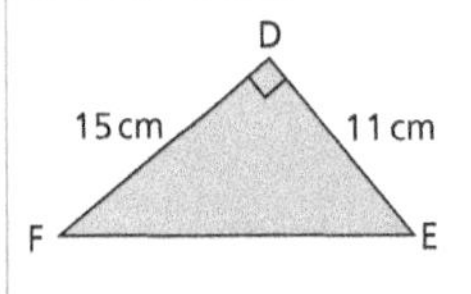

Label the hypotenuse c and the other two sides a and b.

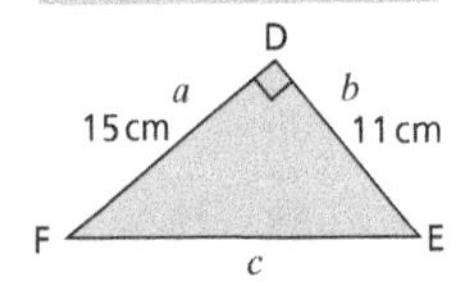

$c^2 = a^2 + b^2$ Write Pythagoras' theorem.

$c^2 = 15^2 + 11^2$ Substitute the values for a and b.

$c^2 = 225 + 121$ Use your calculator.

$c^2 = 346$

$c = \sqrt{346}$

$= 18.601\ 075\ldots$ Square root both sides. Write down the unrounded decimal.

EF = 18.6 cm (1 d.p.) Round to 1 d.p.

Is a triangle right-angled?

Pythagoras' theorem only works in right-angled triangles. If you have a triangle where the sides a, b and c satisfy $c^2 = a^2 + b^2$ then the triangle is right-angled, with hypotenuse c.

a) A triangle has sides 5 cm, 12 cm and 13 cm. Is it a right-angled triangle?

Substitute the lengths into Pythagoras' theorem. Choose the longest length to be c.

$c^2 = a^2 + b^2$

13^2 $5^2 + 12^2$

169 $25 + 144 = 169$

Left-hand side = Right-hand side

The triangle **is** right-angled.

b) A triangle has sides 3 cm, 6 cm and 7 cm. Is it a right-angled triangle?

Substitute the lengths into Pythagoras' theorem. Choose the longest length to be c.

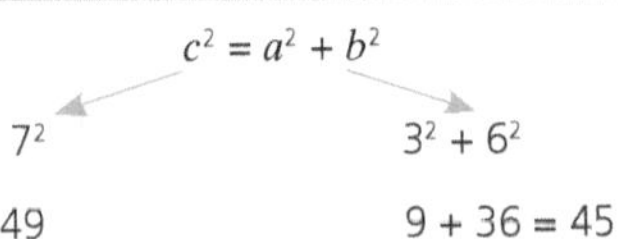

$c^2 = a^2 + b^2$

7^2 $3^2 + 6^2$

49 $9 + 36 = 45$

Left-hand side ≠ Right-hand side

The triangle **is not** right-angled.

RETRIEVE

1 Pythagoras' theorem 1

Finding the hypotenuse

1 Find the length of the hypotenuse in each right-angled triangle.
Round decimal answers to 1 decimal place.

a)

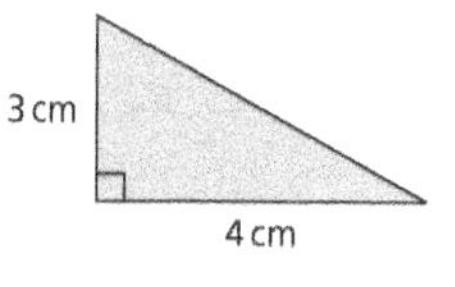

b)

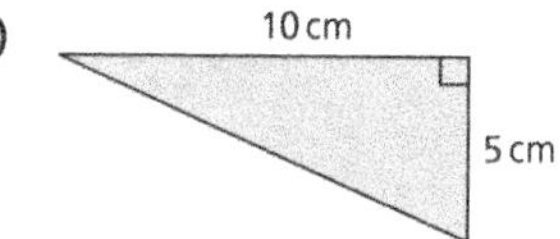

c)

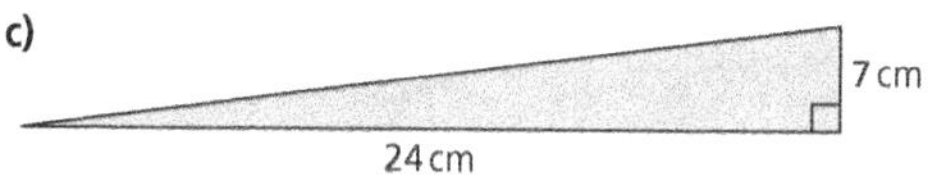

d)

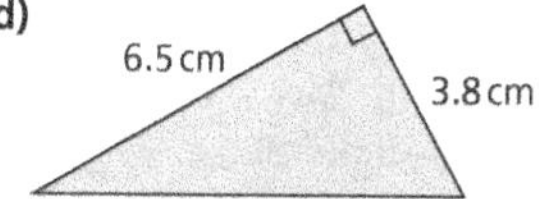

2 Find the length of the hypotenuse in each isosceles right-angled triangle.

a)

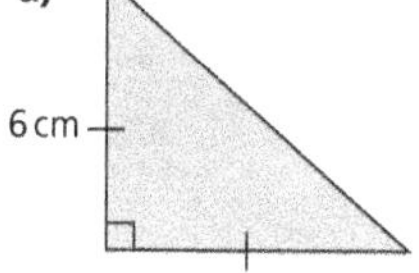

b)

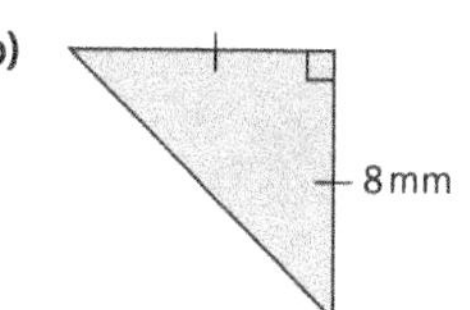

c)

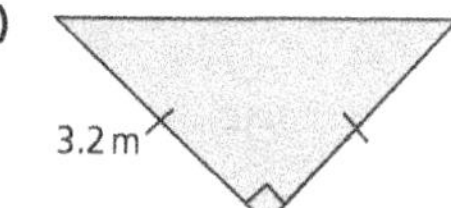

d)

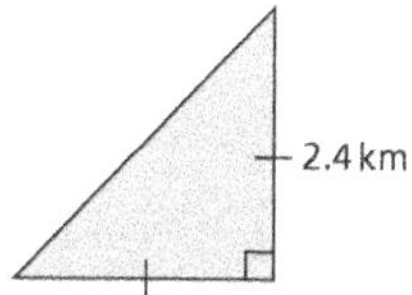

Is a triangle right-angled?

3 A triangle has sides 5 cm, 7 cm and 10 cm.

Is it a right-angled triangle?

4 A triangle has sides 8 cm, 15 cm and 17 cm.

Is it a right-angled triangle?

Pythagoras' theorem 2

Finding one of the shorter sides in a right-angled triangle

You can rearrange $c^2 = a^2 + b^2$ to find either a or b.

$c^2 - a^2 = b^2$ $\qquad$ $c^2 - b^2 = a^2$

$\sqrt{c^2 - a^2} = b$ $\qquad$ $\sqrt{c^2 - b^2} = a$

Or you can substitute the side lengths into $c^2 = a^2 + b^2$ before rearranging.

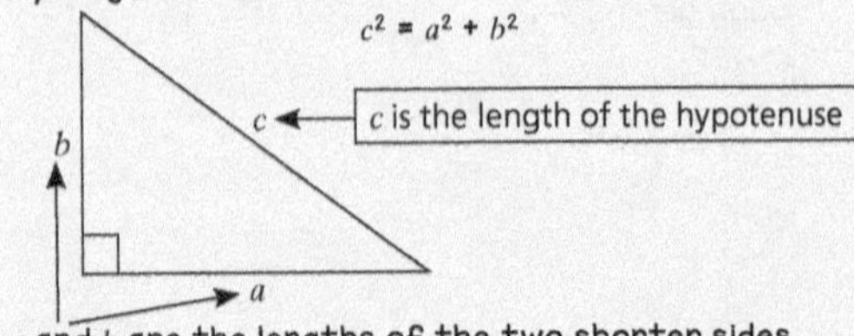

Below is a right-angled triangle ABC.

Find the length of AB. Give your answer to 1 decimal place.

When the answer is a decimal, you may need to round.

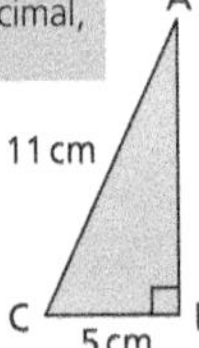

Label the hypotenuse c and the other two sides a and b.

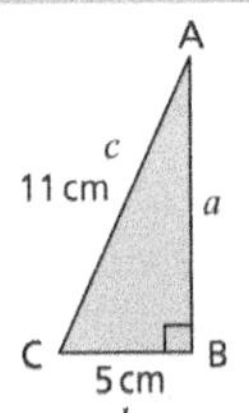

$c^2 = a^2 + b^2$ — Write Pythagoras' theorem.

$11^2 = a^2 + 5^2$ — Substitute the values for b and c.

$121 = a^2 + 25$

$121 - 25 = a^2$ — Subtract 25 from both sides.

$96 = a^2$

$\sqrt{96} = a = 9.797958\ldots$ cm — Square root both sides.

$a = \text{AB} = 9.8$ cm (1 d.p.) — Round to 1 d.p.

Finding one of the shorter sides in a right-angled isosceles triangle

In an isosceles right-angled triangle, $a = b$.

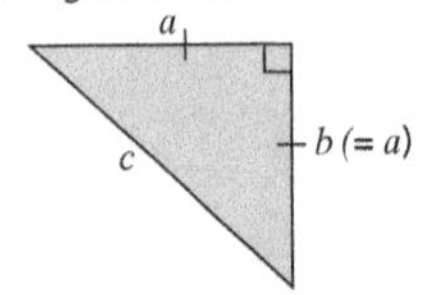

Pythagoras' theorem

$c^2 = a^2 + b^2$

$c^2 = a^2 + a^2 = 2a^2$

Substitute the value of c, then solve to find a.

Below is a right-angled isosceles triangle DEF.

Find the lengths of DE and EF to 1 decimal place.

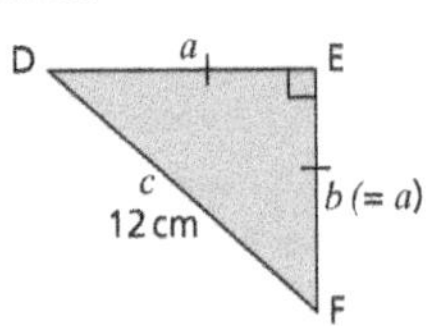

$c^2 = a^2 + b^2$

$12^2 = a^2 + a^2 = 2a^2$ — Substitute the value for c and $b = a$.

$144 = 2a^2$

$72 = a^2$ — Solve to find a.

$a = \sqrt{72} = 8.48528\ldots$ — Square root both sides.

$\text{DE} = \text{EF} = 8.5$ cm (1 d.p.) — Round to 1 d.p.

Pythagorean triples

A **Pythagorean triple** is a set of three positive integers a, b and c, such that $c^2 = a^2 + b^2$.

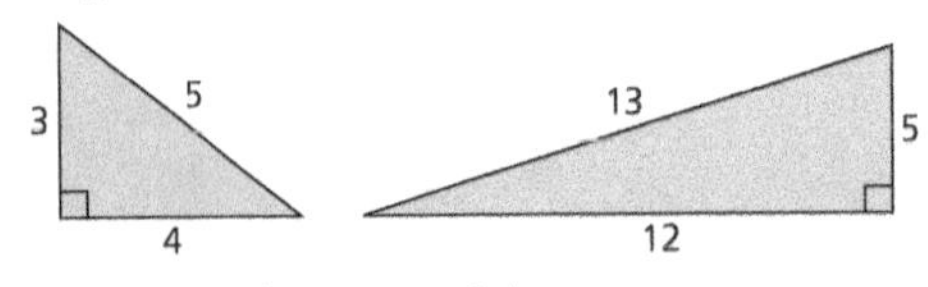

3, 4, 5 is a Pythagorean triple.

5, 12, 13 is a Pythagorean triple.

Decide whether 9, 12, 15 is a Pythagorean triple.

Substitute the values into Pythagoras' theorem. Choose the longest length to be c.

$c^2 = a^2 + b^2$

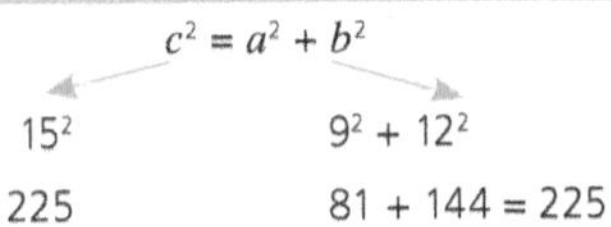

15^2 $\qquad$ $9^2 + 12^2$

225 $\qquad$ $81 + 144 = 225$

Left-hand side = Right-hand side

9, 12, 15 **is** a Pythagorean triple.

RETRIEVE

Pythagoras' theorem 2

Finding one of the shorter sides in a right-angled triangle

1 All these triangles are right-angled.

Find the length of the side labelled with a letter. Round decimal answers to 1 decimal place.

a)

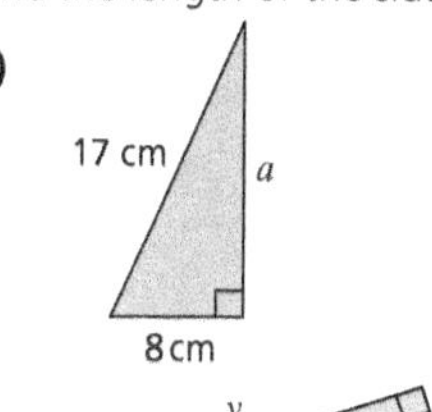

$a =$ ....

b)

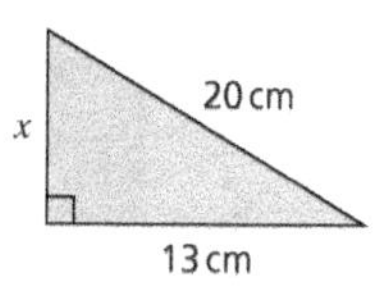

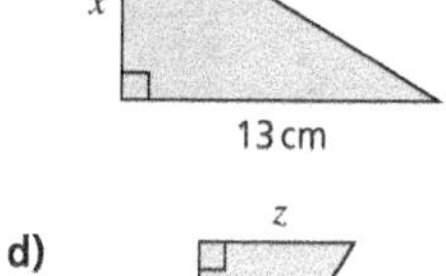

$x =$ ....

c)

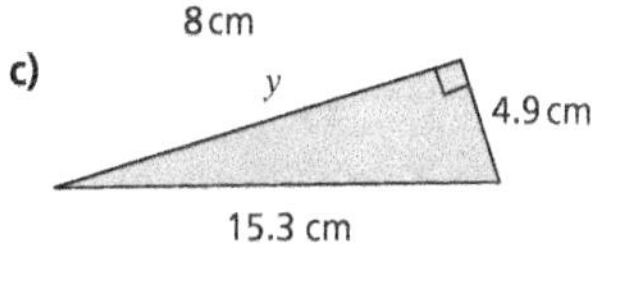

$y =$ ....

d)

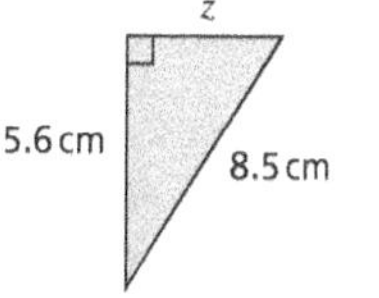

$z =$ ....

2 In a right-angled triangle, the longest side is 85 mm and the shortest side is 36 mm.

Find the length of the other side.

Finding one of the shorter sides in a right-angled isosceles triangle

3 a) Find the length of AC.

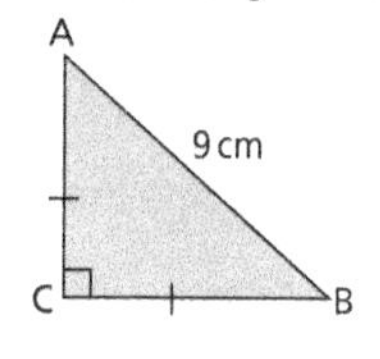

AC = ...

b) Find the length of DE.

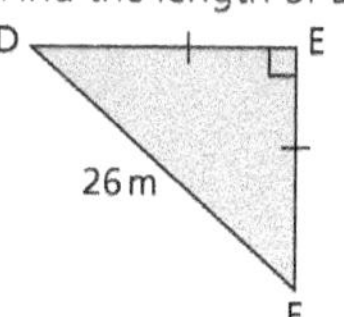

DE = ...

4 The longest side in an isosceles right-angled triangle is 20.4 cm.

Find the lengths of the two shorter sides, to 1 decimal place.

Pythagorean triples

5 Decide whether each set of values is a Pythagorean triple.

a) 7, 10, 15

b) 20, 21, 29

Solving problems using Pythagoras' theorem

Finding a side of a right-angled triangle in real-life situations

Right-angled triangles occur in situations like these:

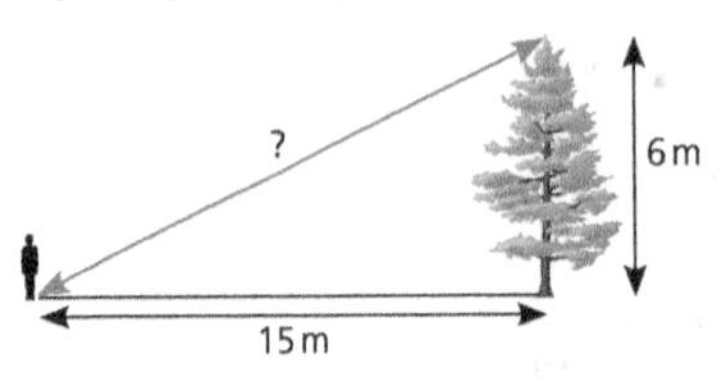

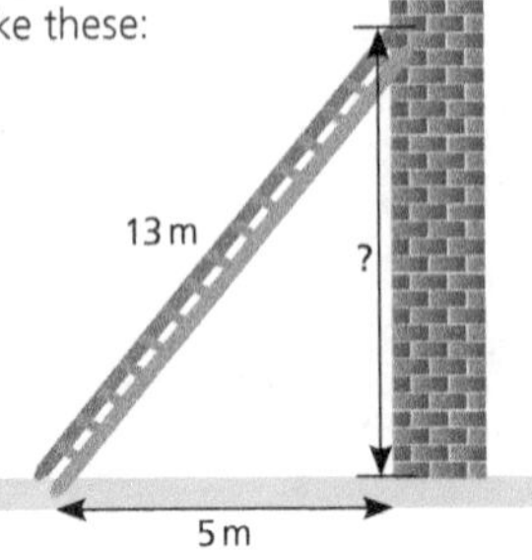

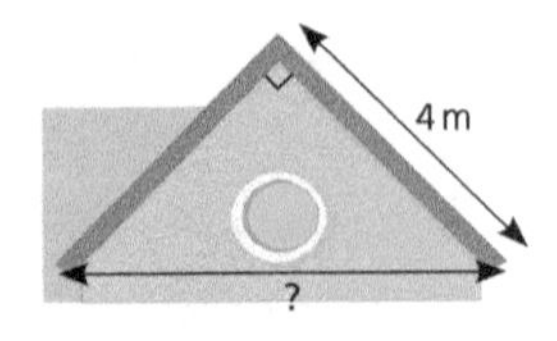

A ladder 20 metres long leans against a wall. The top of the ladder is 11 m above the ground.

Work out the distance of the bottom of the ladder from the wall. Give your answer to the nearest centimetre.

The wall is at right angles to the ground, so this is a right-angled triangle problem.

20 m

11 m

Label the hypotenuse c and the other two sides a and b.

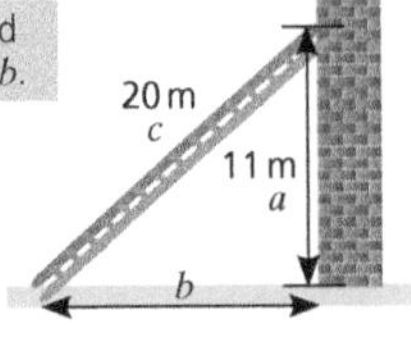

$c^2 = a^2 + b^2$ Write Pythagoras' theorem.

$20^2 = 11^2 + b^2$ Substitute the values for c and a.

$400 = 121 + b^2$

$400 - 121 = b^2$ Subtract 121 from both sides.

$279 = b^2$

$\sqrt{279} = b = 16.7032...$ m Square root both sides.

$= 16.70$ m (to the nearest cm)

The bottom of the ladder is 16.70m from the wall.

Recognising right-angled triangles in other shapes

Right-angled triangles occur in other shapes, such as these:

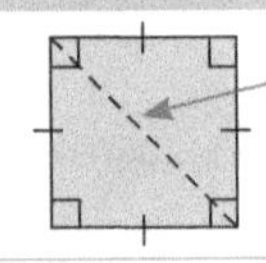

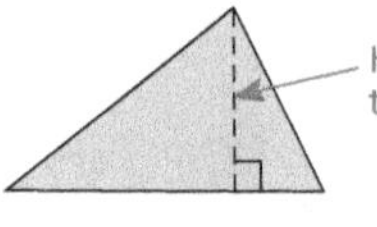

Here is an isosceles triangle.

Work out its area. Give your answer to 2 decimal places.

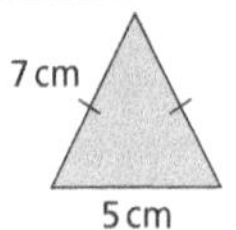

Area of triangle = $\frac{1}{2}bh$

Draw in the height of the triangle

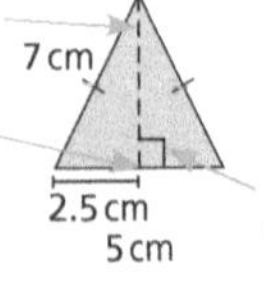

By symmetry, this is the midpoint of the base

Height is always perpendicular (at 90°) to the base

$c^2 = a^2 + b^2$

$7^2 = a^2 + 2.5^2$

$49 - 6.25 = a^2$

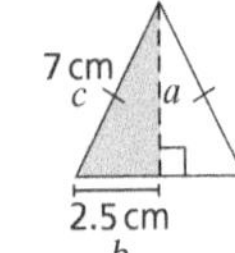

Use Pythagoras' theorem in this half of the triangle to find the height.

$a = \sqrt{42.75} = 6.5383...$ cm Don't round yet.

Area of triangle = $\frac{1}{2}bh$

$= \frac{1}{2} \times 5 \times 6.5383...$

$= 16.34587...\text{cm}^2$

$= 16.35\text{ cm}^2$ (2 d.p.)

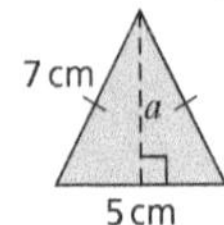

Use the base of the **whole** triangle.

Round the final answer.

Solving problems using Pythagoras' theorem

Finding a side of a right-angled triangle in real-life situations

1. Malia needs a ramp up the step to her front door.

   The height of the step is 0.4 m and the ramp will start 3 m from the step.

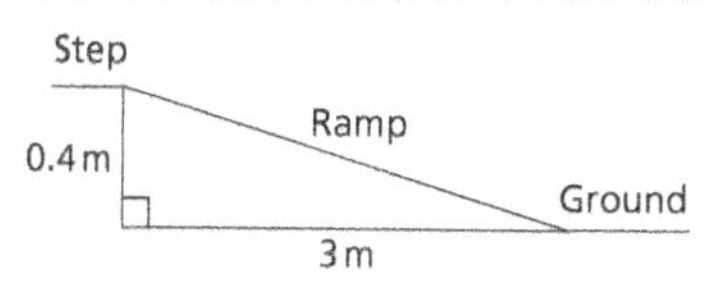

   Work out the length of the ramp, to the nearest centimetre.

2. The size of a TV screen is the length of its diagonal.

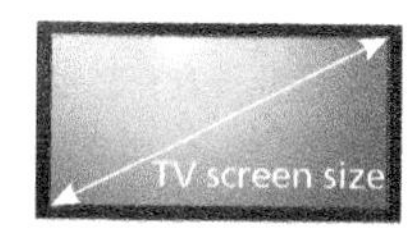

   a) A 42-inch TV screen is a rectangle 32 inches long.

   Work out the height of this TV screen, to the nearest inch.

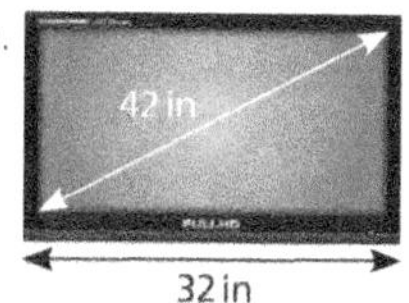

   b) A rectangular TV screen measures 48 inches by 27 inches.

   Calculate the screen size of this TV, to the nearest inch.

3. The diagram shows a 3 m ladder leaning against a wall.

   The bottom of the ladder is 1.2 m from the base of the wall.

   How high up the wall does the ladder reach?
   Give your answer to 2 decimal places.

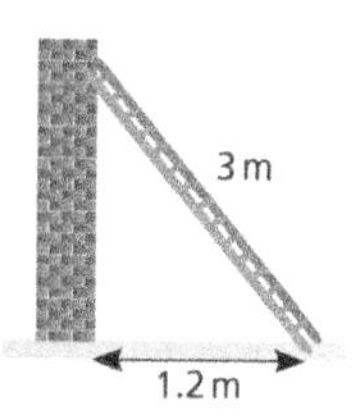

Recognising right-angled triangles in other shapes

4. Find the length of the diagonal of this rectangle.

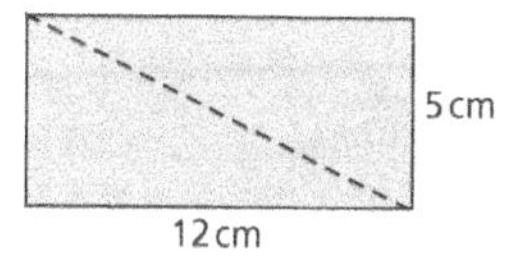

5. a) Work out the height, h, of triangle ABC.

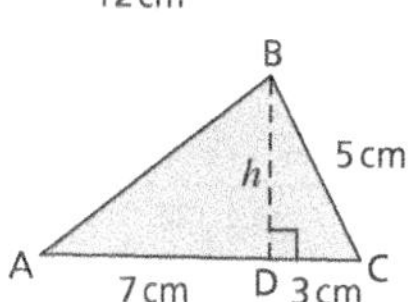

   b) Calculate the area of triangle ABC.

   c) Work out the perimeter of triangle ABC. Give your answer to 1 decimal place.

ORGANISE

2 Introducing probability

Vocabulary of probability

Probability is a measure of how likely something is to happen. It is the likelihood of a particular **event** resulting from an **experiment** or **trial**.

Probability is a measure of the chance of something happening. It can be expressed in words or numbers.

Term	Example (based on rolling dice)
Experiment or **trial** is the procedure you are doing.	Rolling a six-sided dice
The **sample space** is the set of all outcomes (or results) that could occur.	All the numbers you could roll: 1, 2, 3, 4, 5, 6
An **outcome** is the result of a trial.	The specific number that you roll, e.g. a 3.
An **event** is something that can happen in the trial. It is a particular result from a trial and is a subset of the possible outcomes (which can be one outcome or many).	In the event of rolling a 2, the set of outcomes is 2. In the event of rolling an even number, the set of outcomes is 2, 4 and 6.
Mutually exclusive events cannot occur at the same time.	When rolling one dice, you will only roll an even number or an odd number, not both at once.
Independent events are when the probability of the second event does not change based on the first event.	Imagine rolling a dice two times. Whatever you roll on the first roll will not affect what you roll on the second.
Equally likely means the outcomes have the same chance of happening. In this case, the experiment or trial is said to be **fair** (**unbiased**).	The chances of rolling 1, 2, 3, 4, 5 and 6 are the same.
A trial or experiment is **biased** when the outcomes are not equally likely.	A dice that has been weighted so that it is more likely to land on a 6.

Expressing probability in words

Probability can be expressed using words such as **certain**, **likely**, **even chance**, **unlikely**, and **impossible**. Something that is certain will definitely happen. Something that is impossible cannot occur.

Even chance (also called 'evens' or '50/50') means that something has a 50% chance of happening. Note that this is only possible if there are exactly two outcomes that are equally likely. It is not the same as 'equally likely outcomes' when there are more than two outcomes.

The chance of each outcome on a fair coin is $\frac{1}{2}$. This is an 'even chance'.

The chance of spinning each colour is $\frac{1}{3}$. They are equally likely outcomes but not an 'even chance'.

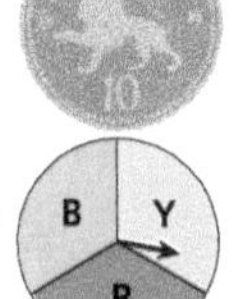

Probability scale

Probability can be measured in numbers using decimals, fractions or percentages on a scale from 0 (impossible) to 1 (certain).

In words	Probability	Example
Certain	1 (or 100%)	New Year's Day will be on 1st January
Likely	Between 0.5 and 1 (50% and 100%)	Pulling out a yellow ball from a bag of 8 yellow balls and 2 green balls
Even chance	0.5 (or $\frac{1}{2}$ or 50%)	Getting a Heads when flipping a fair coin
Unlikely	Between 0 and 0.5 (0% and 50%)	Pulling out a green ball from a bag of 8 yellow balls and 2 green balls
Impossible	0 (or 0%)	Rolling a 7 on a dice numbered 1 to 6

The probability scale can be shown on a number line:

Impossible | Very unlikely | Unlikely | Even chance | Likely | Very likely | Certain

0 (0%) | $\frac{1}{2}$ (50%) | 1 (100%)

2 Introducing probability

Vocabulary of probability

1. Imagine spinning a fair, three-coloured spinner and measuring the probability of spinning red or blue. Draw lines to join the terms to the correct example.

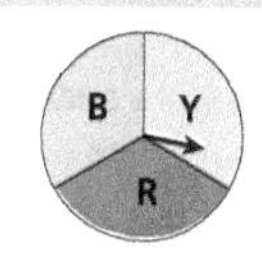

Terms	Example
Event	Spinning the spinner
Outcome	Each section is equal
Sample space	Spinning yellow
Trial	Spinning red or blue
Unbiased	Red, yellow and blue

Expressing probability in words

2. State whether each event is **impossible**, **unlikely**, **even chance**, **likely** or **certain**.

   a) Picking a red card from a standard, fair deck of cards. ....

   b) Rolling a 1 on a fair, 10-sided dice. ....

   c) The next month will have at least 15 days. ....

   d) It will rain meatballs. ....

3. Claire says that there is an even chance of rolling a 6 on a fair, six-sided dice because each outcome is equally likely. Is she correct? Explain your answer.

....

....

Probability scale

4. This probability scale shows the probabilities of five events (A, B, C, D and E).

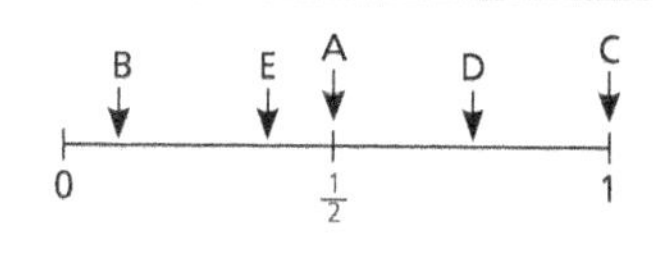

   a) Is event B or E more likely to happen? ....

   b) Which events are less likely to occur than D? ....

   c) Write down whether each event is: **impossible**, **unlikely**, **evens**, **likely** or **certain**. You may need to use the same word for more than one event and all words may not be used.

   A .... B .... C .... D .... E ....

Probability of single events

Calculating theoretical probability

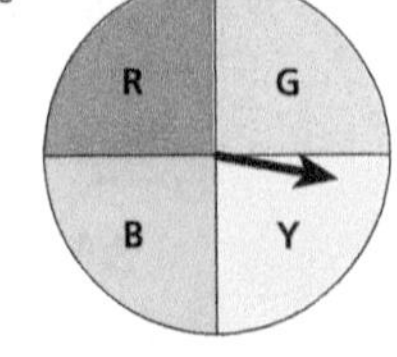

To calculate probability, you need to know all the possible outcomes (also called the **sample space**). The sample space is a set, so set notation S = { } is often used. Outcomes are often written with corresponding letters rather than in words, e.g. B instead of 'blue'. Write the outcomes in an ordered list so that none are missed.

The sample space of flipping a coin is Heads or Tails. This is written S = {H, T}

The sample space of this spinner is Red, Green, Blue, Yellow, or S = {R, G, B, Y}

To calculate probability numerically:

$$\textbf{P(event)} = \frac{\textbf{number of ways the outcome can occur}}{\textbf{total number of possible outcomes}}$$

Find the probability of rolling a number greater than 4 on a fair, six-sided dice.

The sample space is S = {1, 2, 3, 4, 5, 6}. There are six possible outcomes.

The event is rolling a number greater than 4. This can only happen two ways (by rolling a 5 or a 6).

So, P(a number greater than 4) = $\frac{\text{number of ways to roll a number greater than 4}}{\text{total number of ways to roll the dice}} = \frac{2}{6} \left(= \frac{1}{3}\right)$

Probabilities sum to 1

When events are mutually exclusive, they cannot happen at the same time. So the probabilities of all events in a trial of mutually exclusive events will always sum to 1 (or 100%).

Imagine a bag containing 8 yellow marbles (Y) and 2 green marbles (G). One marble is chosen at random.

$P(Y) = \frac{\text{number of yellow marbles}}{\text{total number of marbles}} = \frac{8}{10}$

$P(G) = \frac{\text{number of green marbles}}{\text{total number of marbles}} = \frac{2}{10}$

As fractions, $\frac{8}{10} + \frac{2}{10} = \frac{8+2}{10} = \frac{10}{10} = 1$

As percentages, 80% + 20% = 100%

The total number of marbles will always be the sum of the yellow and green marbles, so the sum of the probabilities will always be a fraction that simplifies to $\frac{1}{1} = 1$.

This fact also means that:

P(event) = 1 – P(event not happening)

You could calculate P(G) by knowing P(Y) and by knowing that all the marbles are green or yellow.

$P(G) = 1 - P(Y) = 1 - \frac{8}{10} = \frac{2}{10}$

Probabilities of mutually exclusive events in a trial will always sum to 1.

Experimental probability

Theoretical probability measures what should happen if everything in the trial is fair.

Experimental probability records the actual results of a trial and reports them as a **relative frequency**.

$$\textbf{Relative frequency} = \frac{\textbf{number of times event occurred}}{\textbf{total number of trials}}$$

The theoretical probability of getting Tails when flipping a coin is 50%. However, if you were to conduct a trial by physically flipping a coin many times, you may find a slightly different chance. Experimental probability will be very close to theoretical probability if a fair trial is repeated enough times.

Before beginning a trial, you can work out the theoretical probability and calculate the **expected results**.

This can be written as:

Expected result = P(event) × number of trials

If you flipped a coin 100 times, you would expect it to land on Tails 50 times.

Expected Tails = $\frac{1}{2} \times 100 = 50$

Suppose a coin is flipped 100 times and comes up Tails 52 times and Heads 48 times.

Relative frequency (T) = $\frac{\text{number of times Tails occurred}}{\text{total number of trials}}$

$= \frac{52}{100} \left(= \frac{13}{25} \text{ or } 0.52\right)$

RETRIEVE

2 Probability of single events

Calculating theoretical probability

1. List the sample space for each trial.

   a) Rolling a six-sided dice

   b) Choosing a letter out of a hat from the word

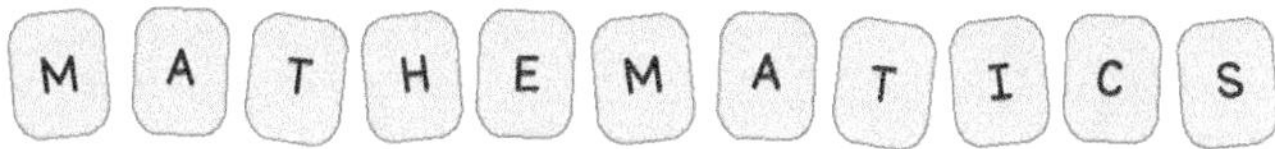

   c) Spinning this spinner

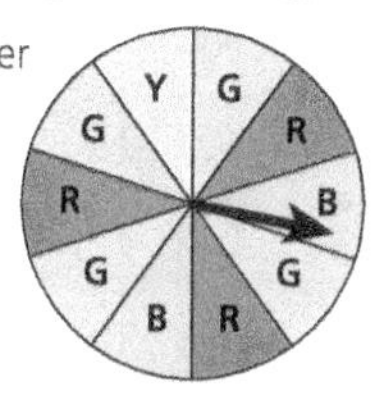

2. Use your answers to question 1 to calculate the probability of:

   a) rolling a number less than or equal to 4

   b) choosing the letter M from the word MATHEMATICS

   c) the spinner landing on green

Probabilities sum to 1

3. A weather forecast says the probability of rain tomorrow is 52%.

   Find the probability that it will **not** rain.

Experimental probability

4. A dice is to be rolled 24 times.

   a) What is the probability of rolling a 6 on a fair dice?

   b) What is the expected number of times a 6 will be rolled?

   c) Here are the results of the trial:

   2 1 6 5 2 5 4 1 2 4 1 1 2 1 3 6 5 4 2 3 4 1 5 3

   What is the relative frequency of rolling a 6?

ORGANISE 2

Calculating probabilities of combined events

Sample spaces of combined events

A **combined event** is when more than one event takes place in a single experiment, e.g. flipping two coins or rolling two dice. To calculate probability in combined events, you need to know all the possible outcomes. You can use a **sample space diagram** to ensure you have considered all the possible outcomes.

Calculating the probability of combined events

Calculating the probability of combined events is the same as for single events.

P(event) = $\frac{\text{number of ways the outcome can occur}}{\text{total number of possible outcomes}}$

The difference is that you must carefully consider the outcomes of a combined event. It is best to start with an organised list or table to find all the possible outcomes. The total number of outcomes will be the product of the number of outcomes of each event.

The outcome of flipping two coins can be shown using a list or a table to show the sample space.

As a list:

H, H
H, T
T, H
T, T

As a table:

1st coin \ 2nd coin	Heads	Tails
Heads	H, H	(H, T)
Tails	(T, H)	T, T

To find the probability of one Head and one Tail when flipping two coins, count the number of ways of flipping HT or TH and divide by the total number of possible outcomes.

There are 2 ways to flip one Head and one Tail and a total of 2 × 2 = 4 possible outcomes.

P(one Head and one Tail) = $\frac{2}{4}\left(=\frac{1}{2}\right)$

The table below shows the sample space for the sum of rolling two dice.

Possible outcomes from second dice

Dice 1 \ Dice 2	1	2	3	4	5	6
1	2	3	4	(5)	6	7
2	3	4	(5)	6	7	8
3	4	(5)	6	7	8	9
4	(5)	6	7	8	9	10
5	6	7	8	9	10	11
6	7	8	9	10	11	12

Possible outcomes from first dice

To find the probability of rolling two dice that sum to 5, count the number of ways to achieve a sum of 5 and divide by the total number of possible outcomes, 6 × 6 = 36.

There are 4 ways to have a sum of 5 out of 36 possible outcomes.

P(sum of 5) = $\frac{4}{36}\left(=\frac{1}{9}\right)$

This spinner is spun and a marble is pulled from the bag.

What is the probability of spinning blue and pulling a blue marble from the bag?

Start by creating a table or diagram showing all the possible outcomes.

Marble \ Spinner	Red	Yellow	Blue	Purple
Blue	BR	BY	(BB)	BP
Blue	BR	BY	(BB)	BP
Red	RR	RY	RB	RP
Red	RR	RY	RB	RP
Red	RR	RY	RB	RP

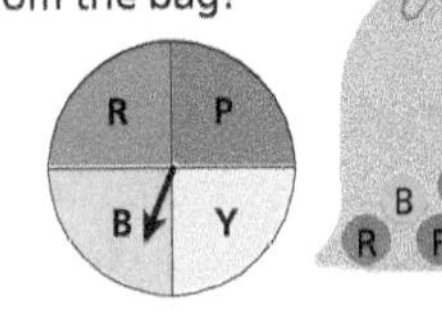

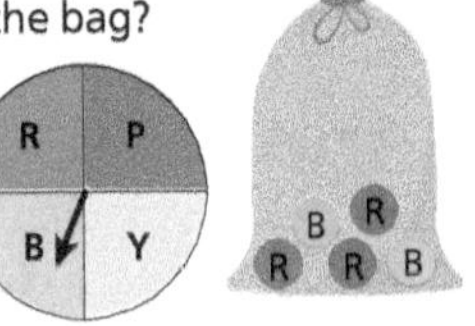

There are 4 × 5 = 20 possible outcomes.

There are 2 ways to spin blue and pull out a blue marble out of 20 possible outcomes.

P(blue spin and blue marble) = $\frac{2}{20}\left(=\frac{1}{10}\right)$

RETRIEVE 2

Calculating probabilities of combined events

Sample spaces of combined events

1 Show the sample spaces for the following.

a) Create a table to show the total possible scores when spinning two spinners numbered 1 to 4.

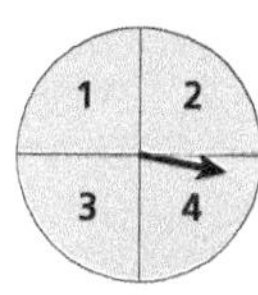

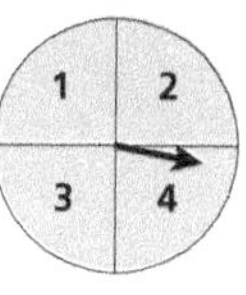

b) Create a list or a table to show the possible outcomes of spinning a four-sided spinner numbered 1 to 4 and flipping a coin.

Calculating the probability of combined events

2 Two four-sided spinners numbered 1 to 4 are spun.

Calculate the probability of:

a) a sum of 4

....

b) a sum of greater than 3

....

3 A four-sided spinner numbered 1 to 4 is spun and a coin is flipped.

Calculate the probability of:

a) an even number and Tails

....

b) **not** an even number and Tails

....

2 Venn diagrams

Understanding Venn diagrams

A **Venn diagram** is a visual way of classifying things, showing what they have in common and how they are different.

The circles represent each set. The information in the overlaps lies in both sets. The information outside of the circles lies in neither set.

Venn diagrams show the relationship between two things when there is something in common between them.

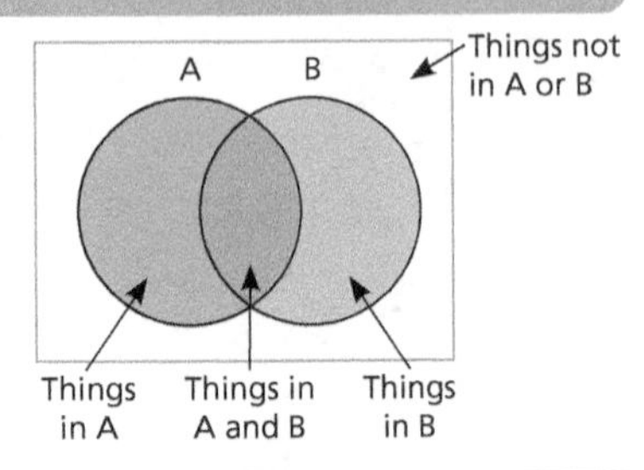

Interpreting and creating Venn diagrams

Venn diagrams are often used to display the results of surveys where people may have more than one answer. For example, what languages people speak or what after-school activities they do.

This Venn diagram shows the results of a survey of 25 children who were asked if they do gymnastics or football after school.

10 children only do gymnastics

8 children do both activities

4 children only do football

3 children do neither activity

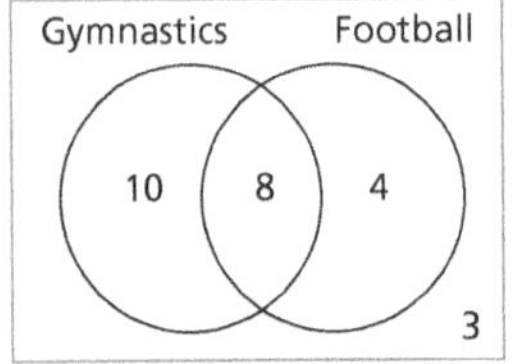

18 children do gymnastics: the 10 that do only gymnastics plus the 8 that do gymnastics and football

12 children do football: the 4 that do only football plus the 8 that do gymnastics and football

This Venn diagram shows whether some students had cereal or eggs for breakfast.

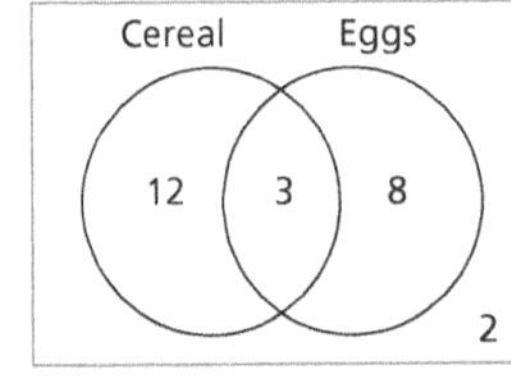

a) How many students were surveyed?

$12 + 3 + 8 + 2 = 25$

25 students were surveyed.

Add up the numbers in the diagram.

b) How many students had eggs for breakfast?

8 students had only eggs plus 3 students had eggs and cereal, so 11 students had eggs.

Create a Venn diagram to represent the following information.

A library recorded whether a set of 50 new books were sci-fi, children's books, or both:

- 32 books were sci-fi.
- 24 books were children's books.
- 21 books were both sci-fi and children's books.
- 15 books were neither sci-fi nor children's books.

Start by filling in the intersection and the outside part. Then calculate the number of books that are only sci-fi and only children's books.

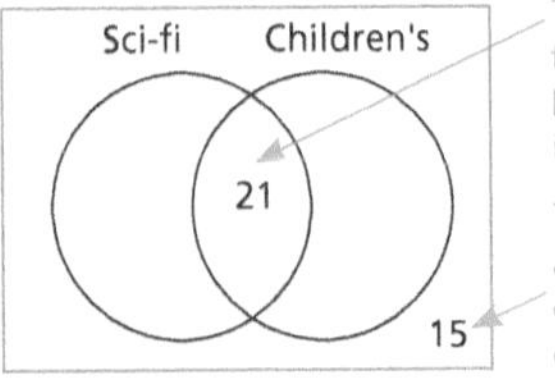

These are 21 books that are sci-fi children's books, so put 21 in the intersection

There are 15 books that are neither sci-fi nor children's, so put 15 outside the circles

There are 32 sci-fi books, including 21 that are children's sci-fi books, so there are $32 - 21 = 11$

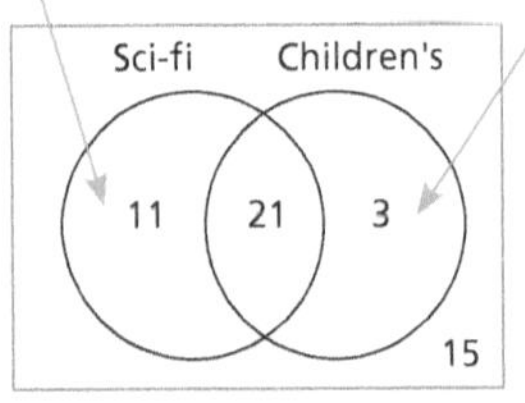

There are 24 children's books, including 21 that are also sci-fi, so $24 - 21 = 3$ that are only children's books

RETRIEVE

2 Venn diagrams

Understanding Venn diagrams

1 What does the shaded area of each Venn diagram represent? Draw lines to join each of the three labels with the correct diagrams.

Set A	Neither Set A nor Set B	Set A and B

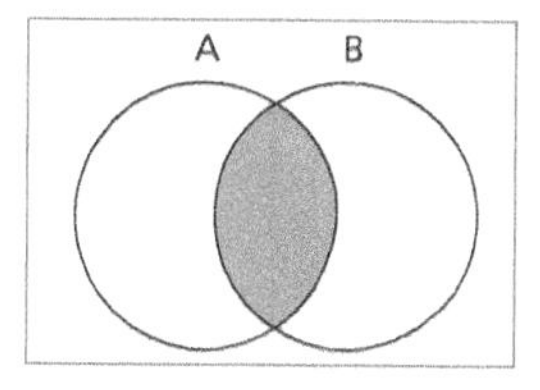

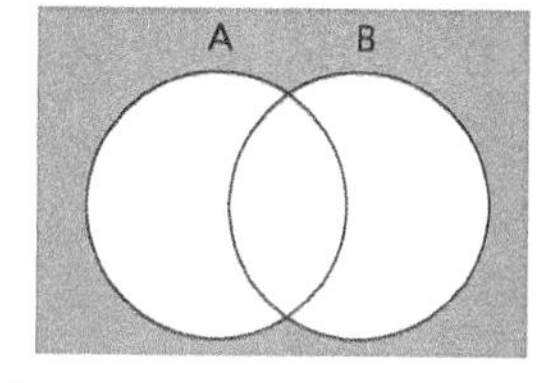

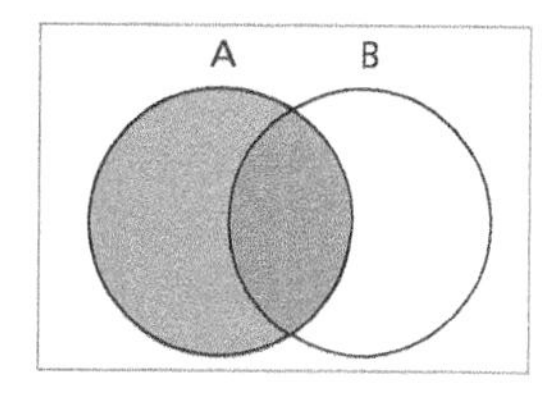

Interpreting and creating Venn diagrams

2 Some workers were asked whether they take the train or the bus to work. This Venn diagram shows the results of the survey.

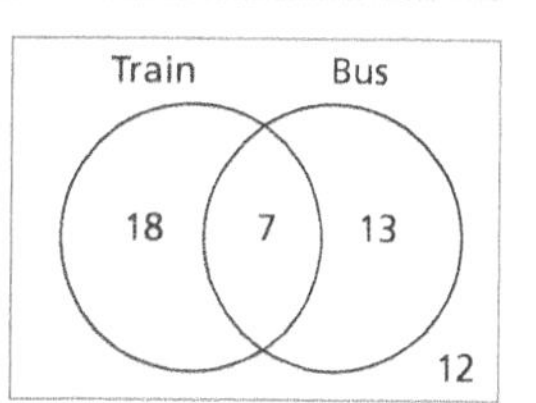

a) How many workers were surveyed? ....

b) How many workers took the train? ....

c) How many workers took the bus and the train? ....

d) How many workers took neither a bus nor a train? ....

3 Complete the Venn diagram using the following information.

60 students were asked whether they like mushroom or sausage on their pizza:

- 7 did not like either topping.
- 18 liked both mushroom and sausage.
- 31 liked sausage.
- 40 liked mushroom.

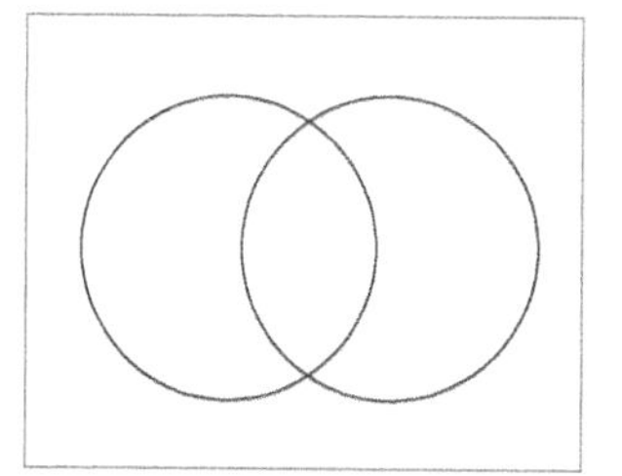

ORGANISE 3

Geometric and special number sequences

Geometric sequences

In a **geometric sequence** each term is multiplied by a **common ratio** to get the next term. To describe a geometric sequence, give the first term and the common ratio.

2, 4, 8, 16, 32, … ($\times 2$ $\times 2$ $\times 2$ $\times 2$) The first term is 2. The common ratio is 2 (as each term is multiplied by 2 to find the next term).

81, 27, 9, 3, 1, … ($\times\frac{1}{3}$ $\times\frac{1}{3}$ $\times\frac{1}{3}$ $\times\frac{1}{3}$) The first term is 81. The common ratio is $\frac{1}{3}$ (as each term is divided by 3 to find the next term).

A famous example of a geometric sequence is Sierpinski's Triangle. In each step, each black triangle is broken up into three smaller black triangles.

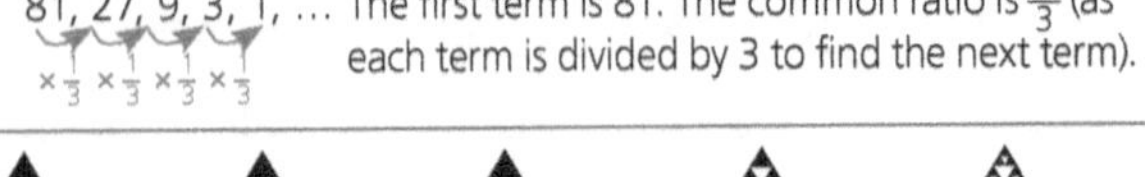

The sequence starts with one black triangle and the number of black triangles is tripled each time, so it has a common ratio of 3.

The first term of a geometric sequence is 2 and the common ratio is 5. Find the first five terms.	1st term is 2 2nd term is $2 \times 5 = 10$ 3rd term is $10 \times 5 = 50$ 4th term is $50 \times 5 = 250$ 5th term is $250 \times 5 = 1250$	The first five terms are 2, 10, 50, 250, 1250, …

In an **arithmetic sequence** the terms increase or decrease by the **same amount** each time. In a **geometric sequence** they increase or decrease by a **common ratio.**

To find the common ratio of a geometric sequence, divide two consecutive terms. It is a good idea to double check by dividing another pair of consecutive terms.

Find the common ratio in the sequence 4, 12, 36, 108, 324, …

$12 \div 4 = 3$ To find the common ratio, divide two consecutive terms.

$36 \div 12 = 3$ To check, divide two other consecutive terms.

The common ratio is 3.

Find the missing term in the sequence 270, 90, ……, 10, $\frac{10}{3}$, …

$90 \div 270 = \frac{1}{3}$

$\frac{10}{3} \div 10 = \frac{1}{3}$

First find the common ratio.

The missing term is $90 \times \frac{1}{3} = 30$

Other number sequences

Square, cube and triangular numbers are sequences that do not have a simple term-to-term rule. In a quadratic sequence, the second difference is constant (i.e. the difference of the differences between the terms is constant).

The sequence of square numbers (1, 4, 9, 16, 25, …) increases by 3, then 5, then 7, then 9, and so on. Each term is generated by squaring the term number.

In the sequence of triangular numbers (1, 3, 6, 10, 15, ...), each term is generated by adding 1 more than the difference between the previous two terms.

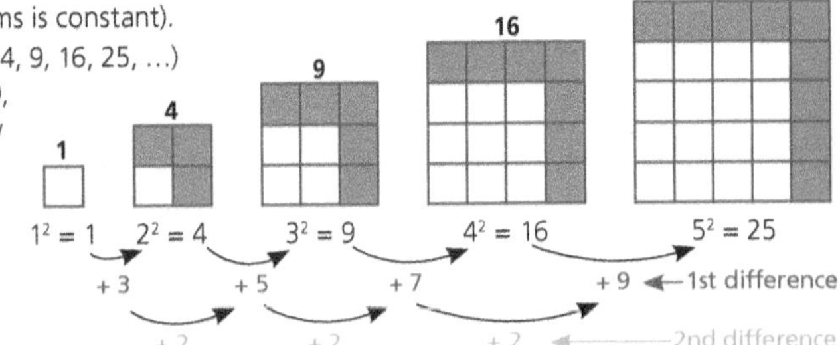

Find the next term in each sequence.

a) 3, 7, 13, 21, 31, ……
$31 + 12 = 43$
The pattern is increasing by sequential even numbers: +4, then +6, then +8, then +10, etc. The next increase will be +12.

b) 70, 55, 43, 34, 28, ……
$28 - 3 = 25$
The pattern is decreasing by multiples of 3: –15, then –12, then –9, then –6, etc. The next decrease will be –3.

You don't always need a term-to-term rule to find the next term of a sequence.

RETRIEVE 3

Geometric and special number sequences

Geometric sequences

1 Write down whether each sequence is geometric or not.

a) 10, 15, 20, 25, 30, ...

....

b) 1500, 300, 60, 12, $\frac{12}{5}$, ...

....

c) 2, 5, 10, 17, 26, ...

....

d) 1, 10, 100, 1000, 10000, ...

....

2 List the first five terms of the geometric sequence with first term 2 and a common ratio of 3.

....

3 Find the missing terms in the following sequences.

a) 2, 8, 32, ..., 512

b) 256, ..., 16, 4, 1

Other number sequences

4 How many dots will be in the next triangle?

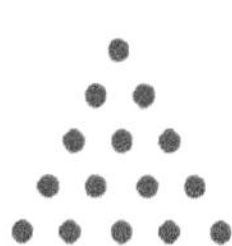

....

5 Find the next term in the sequence 3, 6, 11, 18, 27, ...

ORGANISE

Expanding two or more binomials

The distributive law

The **distributive law** says that the **product** of two numbers multiplied together is the same as the sum of the product of those numbers split into groups, or **partitioned**.

For example, $3(2 + 3)$ can be shown using an area model. Imagine a rectangle split into two sections. To find the area of the whole rectangle, add up the area of each section.

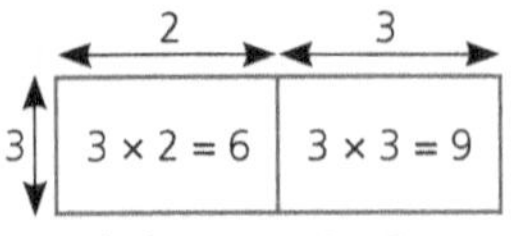

Whole area = 6 + 9
= 15

The distributive law can be shown without visual aids by **expanding the bracket**.

$$3(2 + 3) = (3 \times 2) + (3 \times 3) = 6 + 9 = 15$$

Using an area model (the grid method)

The distributive law also applies when there are variables inside or outside the brackets.

Recall the area method for expanding a single bracket.

	$2x$	3
5	$10x$	15

$5 \times 2x$ 5×3

$5(2x + 3)$
$= (5 \times 2x) + (5 \times 3)$
$= 10x + 15$

Keep any negative sign with the number when partitioning the brackets.

This model can be extended to finding the **product** of two **binomials** (expressions with two terms):

$(2x + 3)(x - 4)$

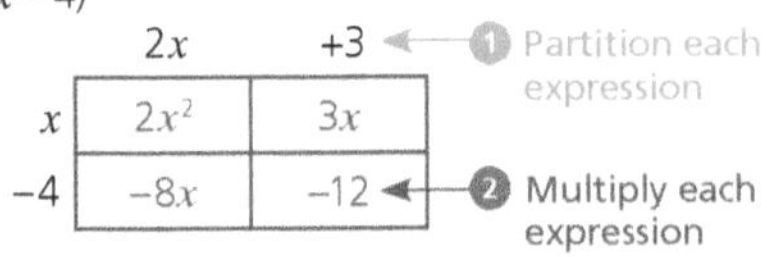

3 Simplify:
$2x^2 + 3x - 8x - 12$
$2x^2 - 5x - 12$
$(2x + 3)(x - 4) = 2x^2 - 5x - 12$

Using the FOIL method

Two brackets can be expanded using a method called FOIL.

First: multiply the first terms of both brackets
Outer: multiply the outer terms of both brackets
Inner: multiply the inner terms of both brackets
Last: multiply the last terms of both brackets

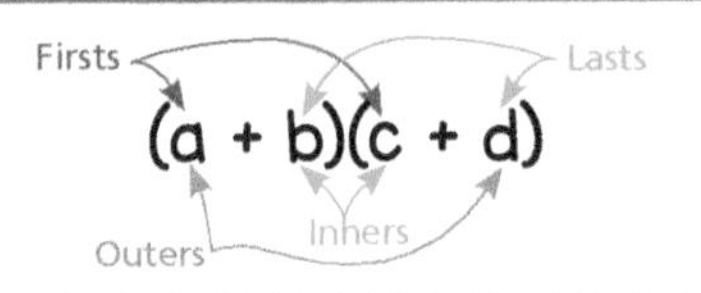

Find the product of $(2x + 3)(x - 4)$

$(2x + 3)(x - 4) = (2x \times x) + (2x \times -4) + (3 \times x) + (3 \times -4)$
$= \quad 2x^2 \quad -8x \quad +3x \quad -12$
$= 2x^2 - 5x - 12$

Make sure you multiply **every** term in the first bracket by **every** term in the second bracket.

Expanding more than two binomials

Both the area model and FOIL can be extended to multiplying more than two brackets. Expand any two brackets first, simplify, then multiply the new expression by the third bracket.

Find the product of $(2x + 3)(x - 4)(x - 3)$

$(2x + 3)(x - 4) = 2x^2 - 5x - 12$ As shown above.

Now multiply $(2x^2 - 5x - 12)(x - 3)$ Multiply the product of the first two brackets by the third bracket.

Area model

	$2x^2$	$-5x$	-12
x	$2x^2 \times x = 2x^3$	$-5x \times x = -5x^2$	$-12 \times x = -12x$
-3	$2x^2 \times -3 = -6x^2$	$-5x \times -3 = 15x$	$-12 \times -3 = 36$

$= 2x^3 - 6x^2 - 5x^2 + 15x - 12x + 36$
$= 2x^3 - 11x^2 + 3x + 36$

FOIL method

$(2x^2 - 5x - 12)(x - 3) =$
$(2x^2 \times x) + (2x^2 \times -3) + (-5x \times x) +$
$(-5x \times -3) + (-12 \times x) + (-12 \times -3)$

You could first multiply $(x - 4)(x - 3)$, then multiply by $(2x + 3)$, and get the same answer.

RETRIEVE

Expanding two or more binomials

The distributive law

1 Find the area of this rectangle using the distributive law.

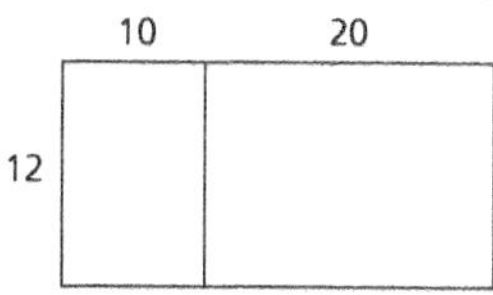

(........ × ....) + (........ × ....) = ... × ....

.... + ... = ...

Using an area model (the grid method)

2 Use an area model to find the product of:

a) $(x - 4)(4x + 3)$

b) $(2n - 1)(3n - 2)$

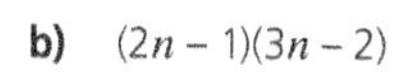

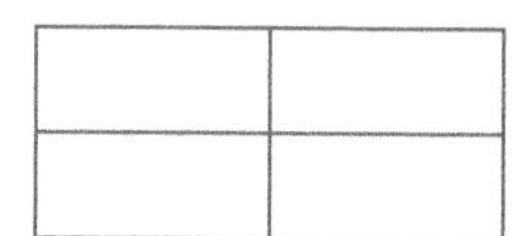

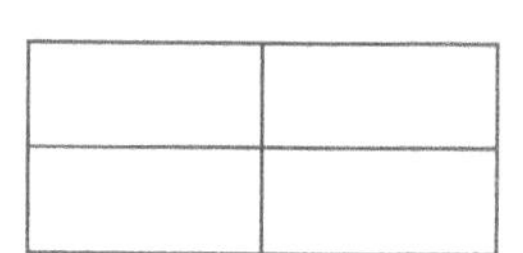

....

Using the FOIL method

3 Use the FOIL method to find the product of:

a) $(4y + 2)(y - 5)$

b) $(2k + 3)(3k + 5)$

....

Expanding more than two binomials

4 Find the product of $(x - 4)(4x + 3)(x - 2)$ using any method. You may wish to use your answer from question 2a) to help you.

....

ORGANISE

Solving multi-step equations 1

Solving one- and two-step equations

Equations are solved using **inverse operations**.

The **one-step equation** $x + 2 = 5$ can be solved as follows:

$x + 2 = 5$

$-2 \quad -2$

$x = 3$

Inverse operations		
Addition +	⇔	Subtraction –
Multiplication ×	⇔	Division ÷
Powers (e.g. x^2, x^3)	⇔	Roots (e.g. $\sqrt{x}$, $\sqrt[3]{x}$)

The process is the same for solving **two-step equations**, except you must first isolate the unknown term (usually the variable x) on one side of the equation. It does not matter which side the unknown term is on.

Solve the equation $3x - 8 = 10$

$3x - 8 = 10$ — Use inverse operations to isolate x. The inverse of -8 is $+8$, so add 8 to both sides.

$+8 \quad +8$

$3x = 18$ — Now use inverse operations to find the value of x. $3x$ means 3 times x, and the inverse of multiplication is division, so divide both sides by 3.

$\div 3 \quad \div 3$

$x = 6$

When solving an equation, carry out the same operation to both sides at each step.

Solving equations with variables on both sides

To solve equations with variables on both sides, isolate the unknown term to one side.

Solve $2x + 8 = 4x - 2$

Subtracting $2x$ from both sides leaves a positive x term, whereas subtracting $4x$ from both sides would leave a negative x term. So it is easier to subtract $2x$.

$2x + 8 = 4x - 2$ — Subtract $2x$ from both sides.

$-2x \quad -2x$

$8 = 2x - 2$ — Use inverse operations to isolate x.

$+2 \quad +2$

$10 = 2x$ — To find the value of x, divide both sides by 2.

$\div 2 \quad \div 2$

$5 = x$ — The answer is usually written with x on the left-hand side.

$x = 5$

Solve $7y - 5 = 5y + 3$

$7y - 5 = 5y + 3$ — Subtract $5y$ from both sides.

$-5y \quad -5y$

$2y - 5 = 3$ — Add 5 to both sides.

$+5 \quad +5$

$2y = 8$ — Divide both sides by 2.

$\div 2 \quad \div 2$

$y = 4$

Solving equations with brackets

When an equation has brackets, first expand them, then simplify and proceed to solve the equation as normal.

Solve $2(3x + 8) - 3(x + 4) = 10$

$2(3x + 8) - 3(x + 4) = 10$ — First expand $2(3x + 8) = 6x + 16$ and $-3(x + 4) = -3x - 12$

$6x + 16 - 3x - 12 = 10$ — Combine like terms.

$3x + 4 = 10$ — The inverse of + 4 is – 4, so subtract 4 from both sides.

$-4 \quad -4$

$3x = 6$ — $3x$ means $3 \times x$, so divide both sides by 3.

$\div 3 \quad \div 3$

$x = 2$

RETRIEVE

Solving multi-step equations 1

Solving one- and two-step equations

1 Solve each equation.

a) $7x + 9 = 51$

$x =$ ....

b) $6x - 9 = 21$

$x =$ ....

Solving equations with variables on both sides

2 Solve each equation.

a) $2x + 8 = 4x - 4$

$x =$ ....

b) $6x - 1 = 15 - 2x$

$x =$ ....

Solving equations with brackets

3 Solve each equation.

a) $3(x - 5) + 2(9 - x) = 10$

$x =$ ....

b) $6x + 7(2x - 3) = 19$

$x =$ ....

Solving multi-step equations 2

Solving equations when the coefficient is negative

When the coefficient of x (or other variable) is negative, manipulate the equation so that the coefficient is positive.

If you have a negative x term in the final step, multiply by −1 to find the value of x.

Solve $10 - 2x = 4$

$10 - 2x = 4$

$+2x \quad +2x$

$10 = 4 + 2x$

$-4 \quad -4$

$6 = 2x$

$\div 2 \quad \div 2$

$x = 3$

Manipulate the equation so that $2x$ is isolated on one side and is a positive value.

The answer is usually written with x on the left-hand side.

Solving equations with one fraction

When solving equations with fractions, remember that a fraction is another way of writing division. For example, $\frac{x}{3}$ means $x \div 3$ and $\frac{4}{x}$ means $4 \div x$.

When an equation has just one fraction, simply multiply both sides by the denominator.

When solving equations with fractions and integers, you can either:

- multiply each term by the denominator of the fraction, **or**
- leave the fraction as it is and multiply at the end.

Solve $\frac{4x - 2}{5} = 2$

$\frac{4x - 2}{\cancel{5}} \times \cancel{5} = 2 \times 5$

$4x - 2 = 10$

$+2 \quad +2$

$4x = 12$

$\div 4 \quad \div 4$

$x = 3$

To 'undo' the fraction, multiply by the denominator.

Solve $4 + \frac{x}{3} = 8$

Method 1: Multiply each term by the denominator first.

$(4 \times 3) + (\frac{x}{3} \times 3) = (8 \times 3)$

$12 + x = 24$

$-12 \quad -12$

$x = 12$

Method 2: Isolate the x term first.

$4 + \frac{x}{3} = 8$

$-4 \quad -4$

$\frac{x}{3} = 4$

$\times 3 \quad \times 3$

$x = 12$

Solving equations with fractions on both sides

When an equation has a fraction on both sides, first find the lowest common multiple (LCM) of the two denominators, then multiply by the LCM to 'undo' them.

Solve $\frac{3x + 1}{2} = \frac{5x + 5}{4}$

$\frac{3x + 1}{2} \times 4 = \frac{5x + 5}{4} \times 4$

$\frac{(3x + 1) \times {}^{2}\cancel{4}}{\cancel{2}} = \frac{(5x + 5) \times \cancel{4}}{\cancel{4}}$

$2(3x + 1) = 5x + 5$

$6x + 2 = 5x + 5$

$-2 \quad -2$

$6x = 5x + 3$

$-5x \quad -5x$

$x = 3$

The LCM of 2 and 4 is 4. Multiply both sides by 4.

Expand the brackets.

RETRIEVE

Solving multi-step equations 2

Solving equations when the coefficient is negative

1 A student is trying to solve the equation $4 - 3x = -2$

The student has made a mistake. Rewrite the solution correctly.

$$4 - 3x = -2$$
$$-4 \qquad\qquad -4$$
$$3x = -6$$
$$x = -2$$

Solving equations with one fraction

2 Solve each equation.

a) $\frac{3k - 2}{4} = 4$

$k =$ ....

b) $2 + \frac{x + 1}{3} = 5$

$x =$ ....

Solving equations with fractions on both sides

3 Solve the equation.

$\frac{3 - 2m}{2} = \frac{m + 2}{6}$

$m =$ ....

Solving inequalities

Inequalities

In an equation, the = sign means that both sides have the same value. In an **inequality**, the sides are **not equal**.

For example, $x < 5$ means x can be any number **less than**, but **not equal to**, 5. x could be 4, 0, −1, or −1.3, etc. It could be any number that is smaller than 5, but not 5.

$x \geqslant 7$ means x can be any number **greater than** or **equal to** 7. x could be 7, 8, 10.5, or 100, etc. It could be 7 or any number greater than 7.

Inequality symbols	
$<$	less than
$\leqslant$	less than or equal to
$>$	greater than
$\geqslant$	greater than or equal to

Double inequalities show that x can take a range of values between two values. For example, $2 < x \leqslant 4$ means that the value of x is between 2 and 4, including 4 but not including 2.

Inequalities on a number line

A number line gives a good visual representation of the possible values of an inequality:

- An empty circle means the inequality is greater than ($>$) or less than ($<$) a certain value.
- A filled circle means the inequality is greater than or equal to ($\geqslant$), or less than or equal to ($\leqslant$), a certain value.
- Arrows are used to show whether the inequality is greater than or less than those values.

$x < 5$:

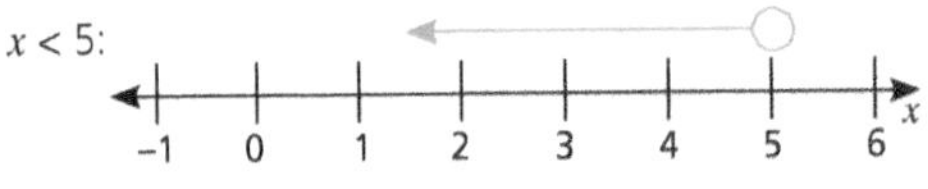

$x \geqslant 7$:

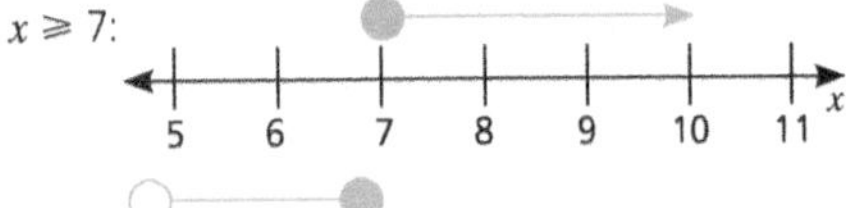

With double inequalities, there will be a line between two circles. $2 < x \leqslant 4$:

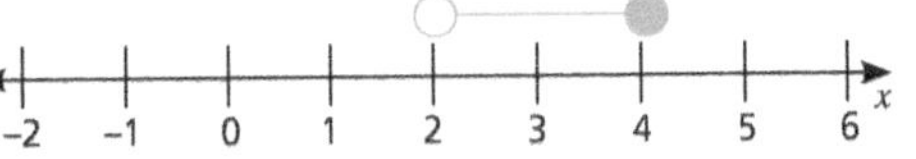

What inequalities are shown by these number lines?

a)

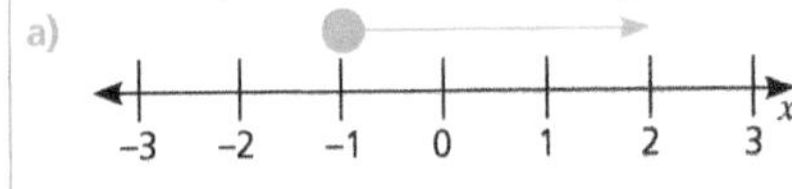

The circle at −1 is filled, so the inequality is 'or equal to'. The arrow is pointing to the numbers greater than −1 so the symbol is $\geqslant$.

The inequality is $x \geqslant -1$.

b)

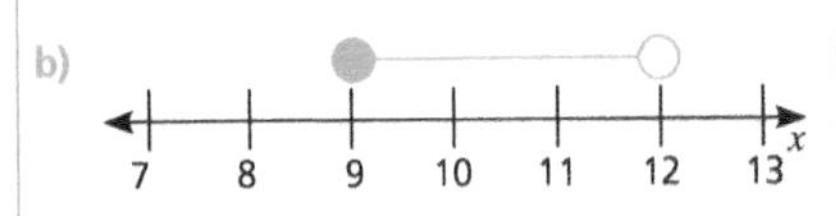

The circle at 9 is filled and the circle at 12 is empty. The line between the two circles means that x is between 9 and 12. The circle at 9 is filled, so $x \geqslant 9$. The circle at 12 is empty, so $x < 12$.

As a double inequality, $9 \leqslant x < 12$.

Solving linear inequalities

The steps for solving inequalities are the same as for equations, except it is important to manipulate the inequality so that the variable (usually x) is positive.

Sometimes a question will also ask you to show the solution on a number line. For $x > -2$:

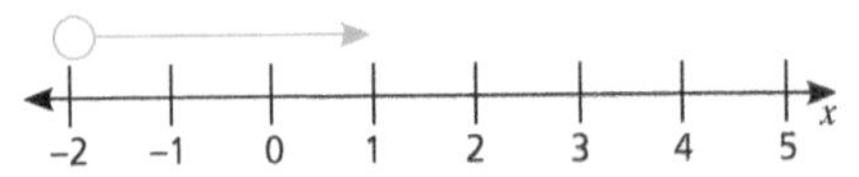

Solve $4 - 2x < 8$

$4 - 2x < 8$

$+2x \quad +2x$

$4 < 8 + 2x$

$-8 \quad -8$

$-4 < 2x$

$\div 2 \quad \div 2$

$-2 < x$

$x > -2$

The answer is usually written with x on the left-hand side. Make sure the inequality stays correct when rewriting the answer. $-2 < x$ means that −2 is less than x or, in other words, x is greater than −2.

RETRIEVE

Solving inequalities

Inequalities

1 For each inequality, write five values that x could be.

a) $x < 7$ ....

b) $x \geqslant 25$ ....

c) $5 \leqslant x \leqslant 10$ ....

Inequalities on a number line

2 Draw each inequality on a number line.

a) $x > -3$

b) $x < 5$

c) $-7 \leqslant x < 2$

Solving linear inequalities

3 Solve each inequality.

a) $4k + 2 < 10$

....

b) $7 - 8m \geqslant -9$

....

c) $4 < 10 - 3j$

....

3 Using and writing formulae

Writing formulae

A **formula** is a rule that shows how two or more **variables** are linked. The variable that is being worked out is called the **subject**.

$A = l \times w$ is a formula. It shows how the variables of length, width and area of a rectangle are related. A is the subject.

A

w

l

For example, a plumber charges a £100 call-out fee plus an additional £25 for every hour worked:

- In words, to calculate the cost, multiply £25 by the number of hours worked and add that to the call-out fee of £100.
- Writing the formula algebraically, state the variables. Let C = the total cost the plumber charges and h = the number of hours worked. Then the formula is $C = 100 + 25h$

Cost (the subject) Call-out fee Price per hour Number of hours (the variable)

A formula shows how two or more variables are related. When writing a formula algebraically, make sure you state what each variable represents.

Using formulae

To use a formula, substitute in the given values of the known variables and find the unknown variable.

A taxi company charges a £5 flat rate plus £1.25 per mile.

a) Write the formula in words.

Multiply the cost per mile, £1.25, by the number of miles travelled and add the product to the flat rate of £5.

b) Write the formula algebraically.

$C = 5 + 1.25m$, where C = the total cost and m = the number of miles travelled

c) Find the cost of taking a taxi for 8 miles.

$C = 5 + 1.25m$

$C = 5 + (1.25 \times 8)$

$C = £15$

To find the cost for travelling 8 miles, substitute $m = 8$.

Speed is the distance travelled per unit of time (e.g. 50 miles per hour means travelling 50 miles in one hour). This relationship can be shown using a formula triangle.

distance
speed time

To use the triangle, cover the subject and multiply or divide the other variables as appropriate.

The triangle shows three formulae:
Speed = distance ÷ time
Time = distance ÷ speed
Distance = speed × time

A plane travels at an average speed of 900 km/h for 8.5 hours. How far has it travelled?

Use the formula triangle, $D = S \times T$, where D = Distance, S = Speed and T = Time.

$D = S \times T$

$D = 900 \times 8.5$

$D = 7650\text{km}$

Substitute $T = 8.5\text{h}$ and $S = 900\text{km/h}$.

The formula $C = \frac{5(F - 32)}{9}$ can be used to convert between degrees Celsius (°C) and degrees Fahrenheit (°F).

Find the temperature in Celsius when it is 85°F, giving the answer to 1 decimal place.

Substitute $F = 85$ into the formula $C = \frac{5(F - 32)}{9}$

$C = \frac{5(85 - 32)}{9}$ First work out the brackets, 85 − 32 = 53

$= \frac{5 \times 53}{9}$ A number written outside brackets means to multiply. 5 × 53 = 265

$= \frac{265}{9}$ A fraction is another way of writing division, so divide. $265 \div 9 = 29.\dot{4}$

= 29.4°C (to 1 d.p.)

RETRIEVE

Using and writing formulae

Writing formulae

1 Write a formula in words and algebraically for each situation.

a) Finding the perimeter of a regular polygon.

In words: ....

Algebraically: ....

b)

In words: ....

Algebraically: ....

Using formulae

2 Use the formulae from question 1 to find the following.

a) The perimeter of a regular decagon (10 sides) with side lengths of 1.3 cm

.... cm

b) The cost of a party with 25 guests

£ ....

3 A train travels 630 km at a speed of 280 km/h.

How many hours does the journey take?

....

ORGANISE

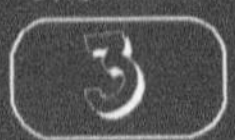

Rearranging formulae

The subject of a formula

The subject of a formula is the variable that is isolated, usually to the left of the equals sign.

Identify the subject in each formula.

a) $A = \frac{b \times h}{2}$

A is the subject. The formula shows how the base and height of a triangle are related to the area.

b) $P = 2w + 2l$

P is the subject. The formula shows how the length and width of a parallelogram are related to the perimeter.

It is sometimes useful to rearrange the formula to **change the subject** to a different variable.

Suppose you know the area and height of a triangle and want to find the base. It would be easier if b was the subject of the formula.

How to rearrange formulae

Rearranging formulae is just like solving equations. Use inverse operations to 'undo' the formula so that a different variable is isolated.

Inverse operations		
Addition +	⇔	Subtraction –
Multiplication ×	⇔	Division ÷
Powers (e.g. x^2, x^3)	⇔	Roots (e.g. $\sqrt{x}$, $\sqrt[3]{x}$)

You can think of formulae as function machines and work backwards to rearrange the subject. To draw a function machine, think about which variable is the input and which variable is the output. To find the input from the output, work backwards using inverse operations.

Function machine:

input → × 3 → + 4 → output

Reverse function machine:

input ← ÷ 3 ← – 4 ← output

Think of rearranging formulae as the same as solving equations. Get the new subject of the formula on its own by using inverse operations.

The formula for the area of a triangle is $A = \frac{bh}{2}$. To rearrange the formula to make the base, b, the subject, think about the operations that are required and apply the inverses.

base → × height → ÷ 2 → area

base ← ÷ height ← × 2 ← area

To show this algebraically, think about undoing each operation to isolate b.

$A = \frac{bh}{2}$ — Multiply both sides by 2 to 'undo' the fraction.

$A \times 2 = \frac{bh}{2} \times 2$ — Simplify.

$2A = bh$

$\div h \quad \div h$ — Divide both sides by h to solve for b.

$2A \div h = b$

$b = \frac{2A}{h}$ — Rewrite the formula to have the b term (the new subject) on the left-hand side.

Rearrange the formula $V = \pi r^2 h$ to make r the subject.

Using a function machine:

r → squared → × π → × h → V

Reverse function machine:

r ← square root ← ÷ π ← ÷ h ← V

Algebraically: $V = \pi r^2 h$ — Divide both sides by πh to isolate the r term.

$\div \pi h \quad \div \pi h$

$\frac{V}{\pi h} = r^2$ — Take the square root of both sides to 'undo' r^2.

$\sqrt{\frac{V}{\pi h}} = \sqrt{r^2}$ — Rewrite the formula to have the r term on the left-hand side.

$r = \sqrt{\frac{V}{\pi h}}$

RETRIEVE

3 Rearranging formulae

The subject of a formula

1 Write down the subject of each formula.

a) $P = sl$ ....

b) $A = \pi r^2$ ....

c) $C = 30h + 45$ ....

How to rearrange formulae

2 You are given the formula $P = sl$, where P is the perimeter of a regular polygon, s is the number of sides and l is the side length.

a) Rearrange the formula to make l the subject.

....

b) Find the length of each side of a 12-sided polygon with perimeter of 108 cm.

....

3 You are given the formula $A = \pi r^2$, where A is the area of a circle and r is the radius.

a) Rearrange the formula to make r the subject.

....

b) Find the radius of a circle with area 225π cm^2.

....

4 You are given the formula $C = 30h + 45$, where C is the charge in £ for a DJ and h is the number of hours worked by the DJ.

a) Rearrange the formula to make h the subject.

....

b) Find the number of hours worked by the DJ if the charge was £135.

....

4 Trigonometry 1

Using the sin ratio to find the length of a side

The hypotenuse of a right-angled triangle is:

- the longest side
- opposite the right angle.

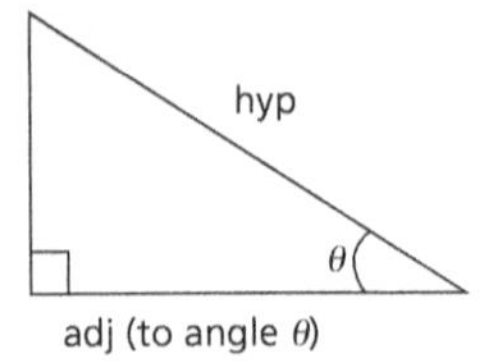

Sin ratio:

$\sin\theta = \dfrac{\text{opposite}}{\text{hypotenuse}} = \dfrac{\text{opp}}{\text{hyp}}$

Here is a right-angled triangle, ABC.
Angle BAC = 42°

Find the length of x to 1 decimal place.

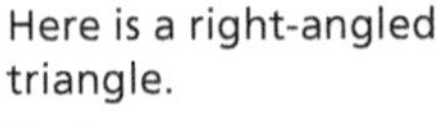

Label the hypotenuse, opposite and adjacent.

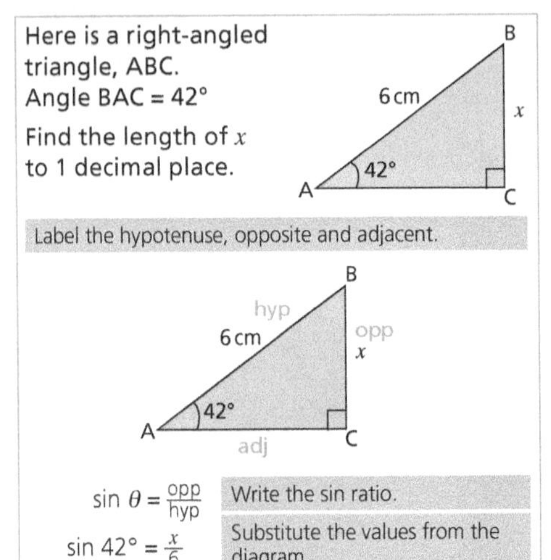

$\sin\theta = \dfrac{\text{opp}}{\text{hyp}}$	Write the sin ratio.
$\sin 42° = \dfrac{x}{6}$	Substitute the values from the diagram.
$6 \times \sin 42° = x$	Multiply both sides by 6.
$4.01478... = x$	Use your calculator.

$x = 4.0$ cm (1 d.p.)

Here is a right-angled triangle.

Find x.

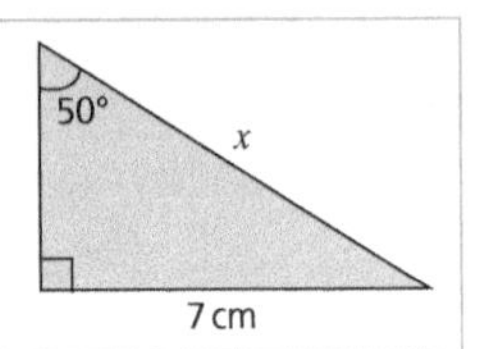

Label the hypotenuse, opposite and adjacent.

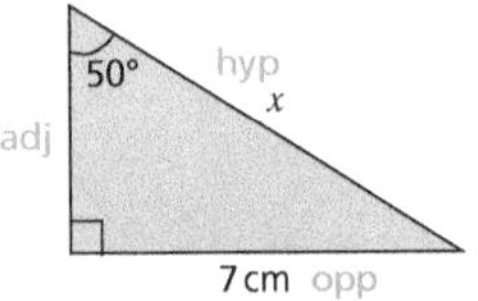

$\sin\theta = \dfrac{\text{opp}}{\text{hyp}}$	You know the opp and you want to find the hyp, so use the sin ratio.
$\sin 50° = \dfrac{7}{x}$	
$x \sin 50° = 7$	Multiply both sides by x.
$x = \dfrac{7}{\sin 50°}$	Divide both sides by sin 50°.
$x = 9.13785...$	Use your calculator.

$x = 9.1$ cm (1 d.p.)

Using the sin ratio to find an angle

When you know the value of sin θ, use $\sin^{-1}$ to find the value of θ. For example:

$\sin\theta = \frac{2}{3}$ $\quad\theta = \sin^{-1}\left(\frac{2}{3}\right)$ Use the $\sin^{-1}$ function on your calculator (usually above the sin key).

ABC is a right-angled triangle.

Find the size of θ.

A, B, C, θ, 10 cm, 6 cm

Label the hypotenuse, opposite and adjacent.

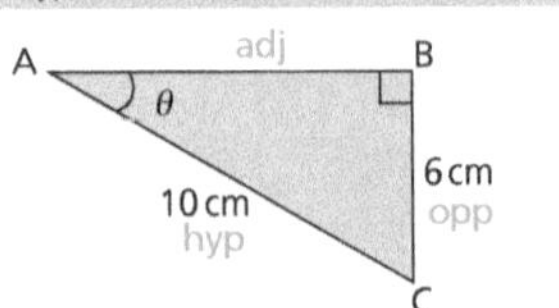

$\sin\theta = \dfrac{\text{opp}}{\text{hyp}}$	You know the opp and hyp, so use the sin ratio.
$\sin\theta = \dfrac{6}{10}$	Substitute the values from the diagram. You don't need to simplify $\frac{6}{10}$; you are going to put it into your calculator.
$\theta = \sin^{-1}\left(\dfrac{6}{10}\right)$	Use $\sin^{-1}$
$\theta = 36.8698...°$	Use your calculator.
$\theta = 36.9°$ (1 d.p.)	1 decimal place is a suitable degree of accuracy for angle measurements.

Trigonometry 1

Using the sin ratio to find the length of a side

1 All these triangles are right-angled.

Find the length of the side labelled with a letter. Round decimal answers to 1 decimal place.

a)

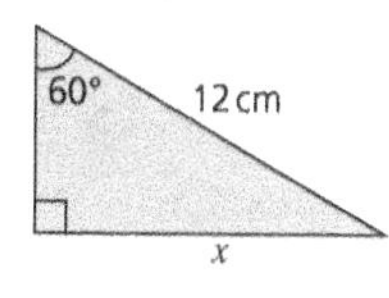

$x =$ ....

b)

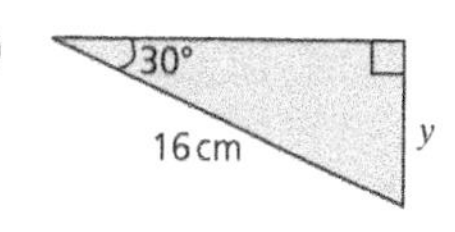

$y =$ ....

c)

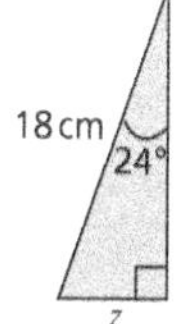

$z =$ ....

d)

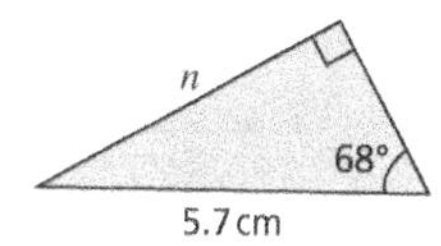

$n =$ ....

2 Both these triangles are right-angled.

Find the length of the side labelled with a letter. Round decimal answers to 1 decimal place.

a)

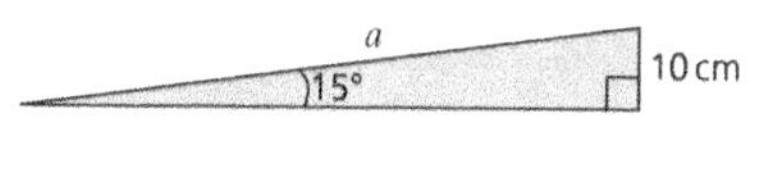

$a =$ ....

b)

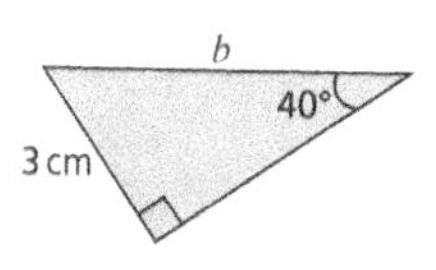

$b =$ ....

3 Here is a right-angled isosceles triangle.

10 cm

a) Use angle facts to find and label the sizes of its angles. .... and ....

b) Use the sin ratio to find the lengths of its shorter sides. .... and ....

Using the sin ratio to find an angle

4 Find the size of the angle labelled θ in each right-angled triangle.

a)

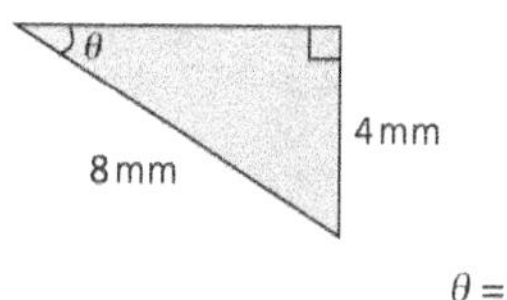

$\theta =$ ....

b)

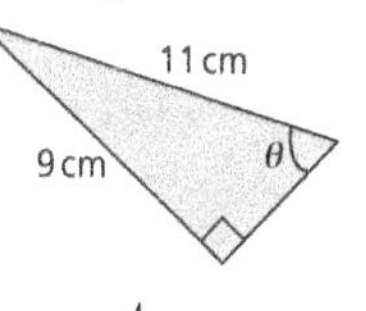

$\theta =$ ....

c)

3 cm, 5 cm, 4 cm, θ

$\theta =$ ....

d)

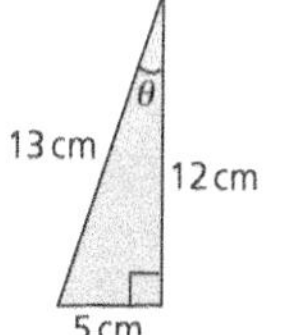

$\theta =$ ....

ORGANISE

4 Trigonometry 2

Using the cos and tan ratios to find the length of a side

Cos ratio:

$\cos\theta = \frac{\text{adjacent}}{\text{hypotenuse}} = \frac{\text{adj}}{\text{hyp}}$

Tan ratio:

$\tan\theta = \frac{\text{opposite}}{\text{adjacent}} = \frac{\text{opp}}{\text{adj}}$

You can use SOHCAHTOA to help you remember the sin, cos and tan ratios.

S O H C A H T O A

$\sin = \frac{\text{opp}}{\text{hyp}}$ $\cos = \frac{\text{adj}}{\text{hyp}}$ $\tan = \frac{\text{opp}}{\text{adj}}$

Sin, cos and tan are abbreviations for sine, cosine and tangent.

Here is a right-angled triangle, DEF.

Angle DEF = 48°

Find the length of x to 1 decimal place.

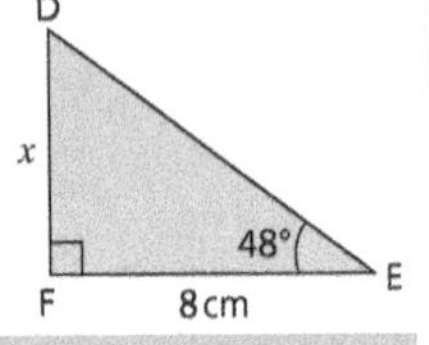

Label the hypotenuse, opposite and adjacent.

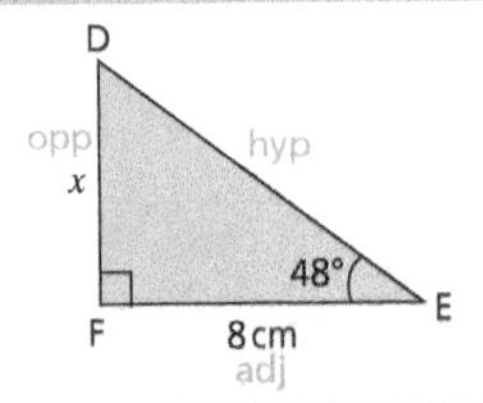

$\tan = \frac{\text{opp}}{\text{adj}}$ — You know the adj and you want to find the opp, so use the tan ratio.

$\tan 48° = \frac{x}{8}$ — Substitute the values from the diagram.

$8 \times \tan 48° = x$ — Multiply both sides by 8.

$8.884900\ldots = x$ — Use your calculator.

$x = 8.9\text{ cm (1 d.p.)}$

Here is a right-angled triangle.

Find x.

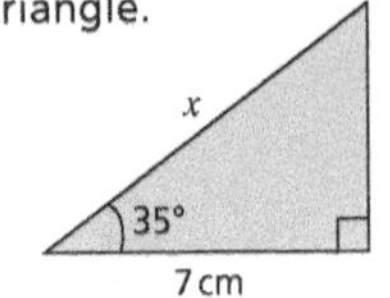

Label the hypotenuse, opposite and adjacent.

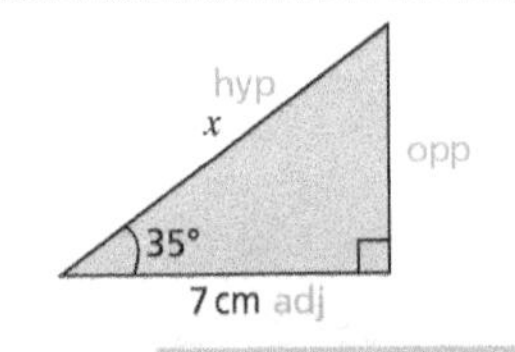

$\cos = \frac{\text{adj}}{\text{hyp}}$ — You know the adj and you want to find the hyp, so use the cos ratio.

$\cos 35° = \frac{7}{x}$

$x \cos 35° = 7$ — Multiply both sides by x.

$x = \frac{7}{\cos 35°}$ — Divide both sides by cos 35°

$x = 8.54542\ldots$ — Use your calculator.

$x = 8.5\text{ cm (1 d.p.)}$

Using the cos and tan ratios to find an angle

When you know the value of cos θ, use $\cos^{-1}$ to find the value of θ. For example:

$\cos\theta = \frac{2}{3}$

$\theta = \cos^{-1}\left(\frac{2}{3}\right)$

When you know the value of tan θ, use $\tan^{-1}$ to find the value of θ. For example:

$\tan\theta = \frac{2}{3}$

$\theta = \tan^{-1}\left(\frac{2}{3}\right)$

Use the $\cos^{-1}$ and $\tan^{-1}$ functions on your calculator (usually above the cos and tan keys).

Here is a right-angled triangle.

Find the size of θ.

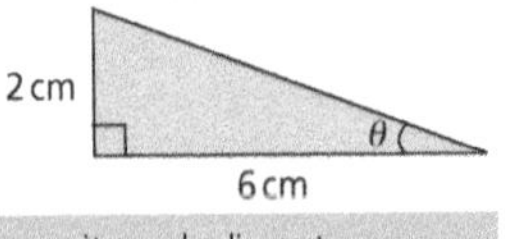

Label the hypotenuse, opposite and adjacent.

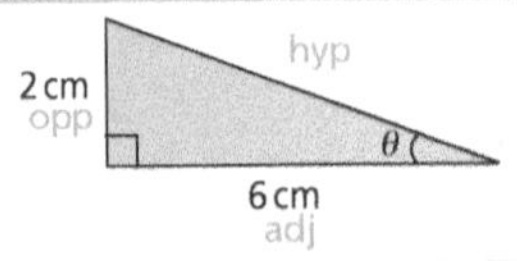

$\tan = \frac{\text{opp}}{\text{adj}}$ — You know the opp and adj so use the tan ratio.

$\tan\theta = \frac{2}{6}$ — Substitute the values from the diagram. You don't need to simplify $\frac{2}{6}$; you are going to put it into your calculator.

$\theta = \tan^{-1}\left(\frac{2}{6}\right)$ — Use $\tan^{-1}$

$\theta = 18.4349\ldots°$ — Use your calculator.

$\theta = 18.4°\text{ (1 d.p.)}$ — 1 decimal place is a suitable degree of accuracy for angle measurements.

RETRIEVE

4 Trigonometry 2

Using the cos and tan ratios to find the length of a side

1. Both these triangles are right-angled.

   Use the cos ratio to find the length of the side labelled with a letter.
   Round decimal answers to 1 decimal place.

   a)
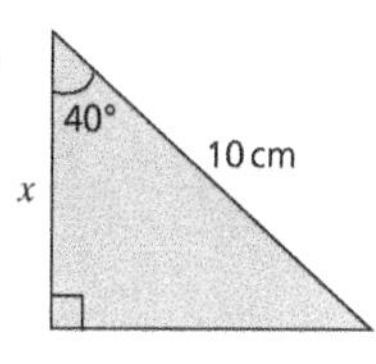

   $x =$ ....

   b)
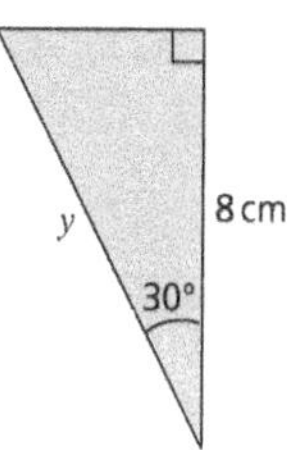

   $y =$ ....

2. Both these triangles are right-angled.

   Use the tan ratio to find the length of the side labelled with a letter.
   Round decimal answers to 1 decimal place.

   a)
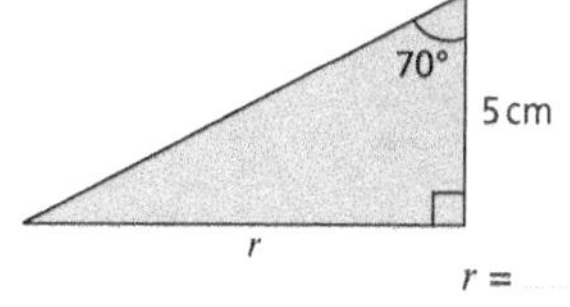

   $r =$ ....

   b)
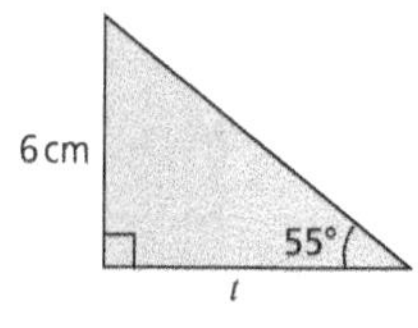

   $t =$ ....

3. Both these triangles are right-angled.

   Find the length of the side labelled with a letter. Round decimal answers to 1 decimal place.

   a)
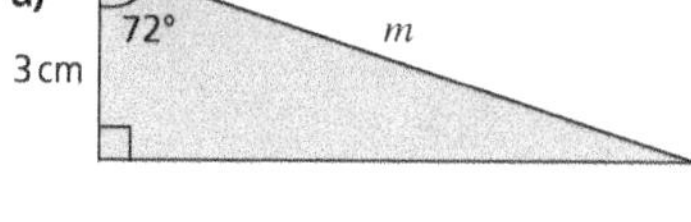

   $m =$ ....

   b)
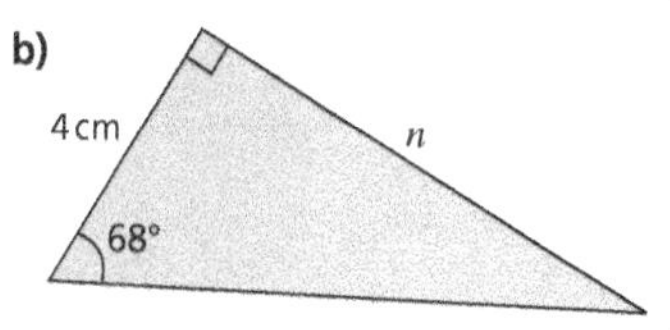

   $n =$ ....

Using the cos and tan ratios to find an angle

4. Find the size of the angle labelled θ in each right-angled triangle.

   a)
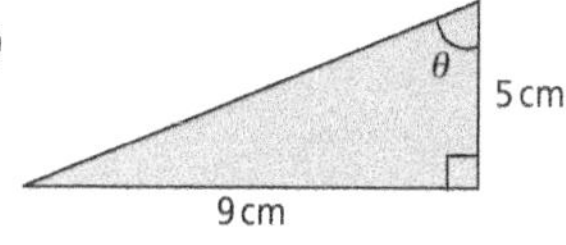

   $\theta =$ ....

   b)
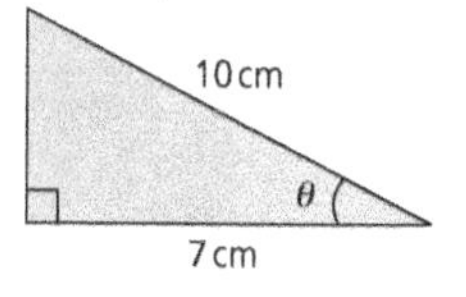

   $\theta =$ ....

   c)
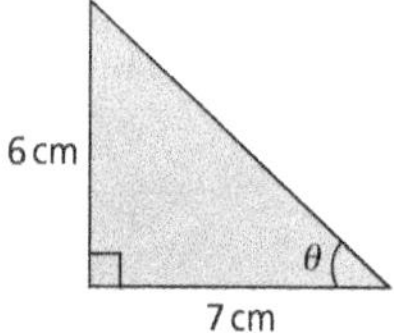
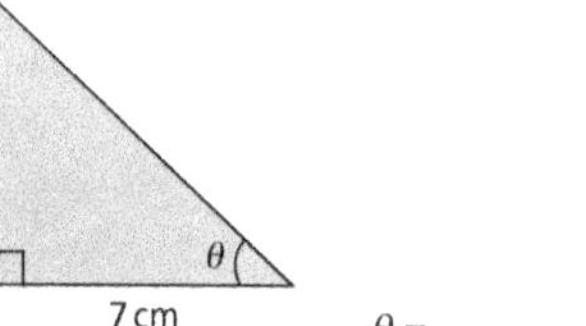

   $\theta =$ ....

   d)
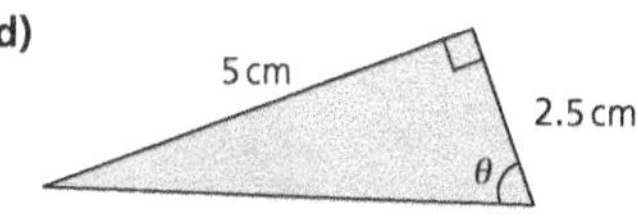

   $\theta =$ ....

ORGANISE

Solving problems involving trigonometry

Finding a side of a right-angled triangle in real-life situations

Right-angled triangles occur in situations like these:

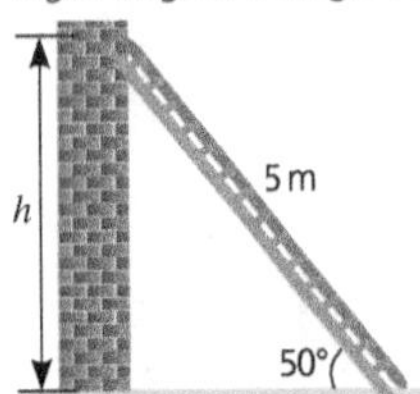

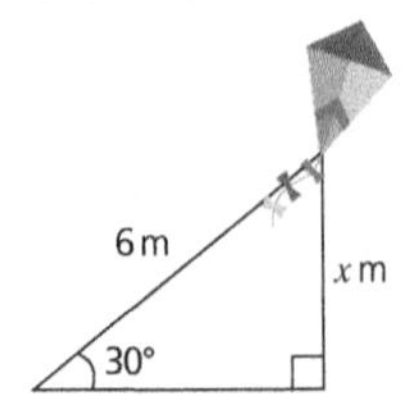

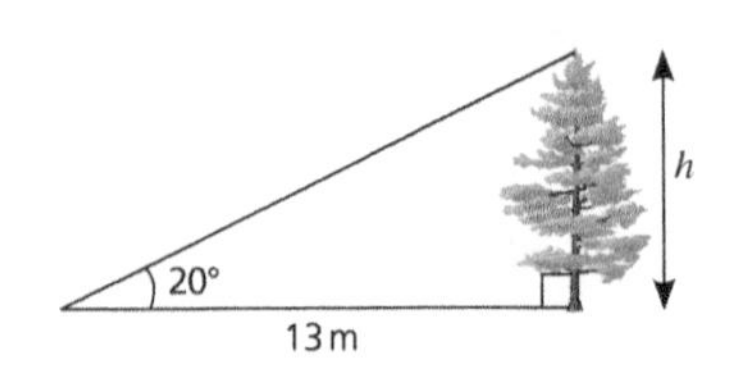

A 4-metre ladder leans against a wall. The angle between the ladder and the ground is 70°. How far up the wall does the ladder reach? Give your answer to the nearest centimetre.

Draw a diagram and label the sides.

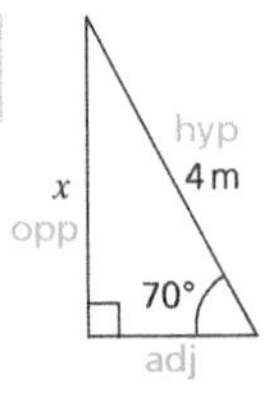

$\sin\theta = \frac{\text{opp}}{\text{hyp}}$ — You know the hyp and you want to find the opp, so use the sin ratio.

$\sin 70° = \frac{x}{4}$ — Substitute the values from the diagram.

$4 \times \sin 70° = x$ — Solve for x.

$x = 3.75877\ldots\text{m}$

$x = 3.76\text{m}$ (to the nearest cm)

Recognising right-angled triangles in other shapes

Right-angled triangles occur in other shapes like these:

Diagonals of squares and rectangles

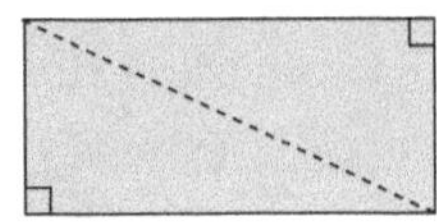

Heights of triangles

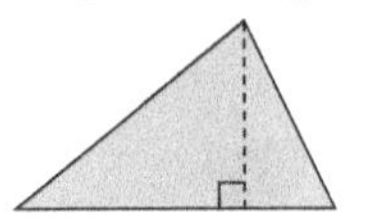

Here is triangle ABC.

a) Find the height, h, of triangle ABC.

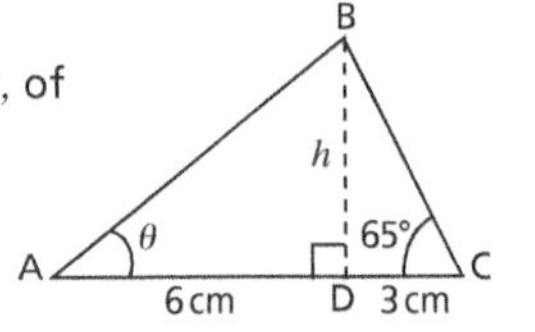

To find h you need the right-angled triangle that includes h, an angle and a side length.

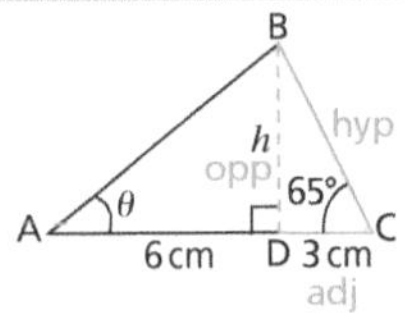

$\tan = \frac{\text{opp}}{\text{adj}}$ — You know the adj and you want to find the opp, so use the tan ratio.

$\tan 65° = \frac{h}{3}$ — Substitute.

$3 \times \tan 65° = h$ — Solve.

$h = 6.4335\ldots$

$h = 6.4\text{cm}$ (1 d.p.) — 1 decimal place is a suitable level of accuracy for measurements in centimetres.

b) Find θ.

To find θ you need the right-angled triangle that includes two side lengths.

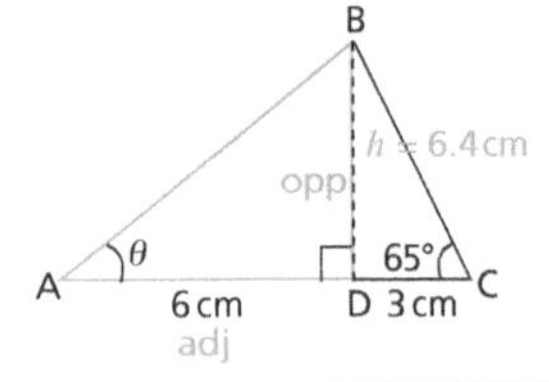

$\tan = \frac{\text{opp}}{\text{adj}}$ — You know the opp and the adj, so use the tan ratio.

$\tan\theta = \frac{6.4}{6}$ — Substitute.

$\theta = \tan^{-1}\left(\frac{6.4}{6}\right)$ — Use $\tan^{-1}$

$\theta = 46.8476\ldots°$

$\theta = 46.8°$ (1 d.p.) — 1 decimal place is a suitable level of accuracy for measurements in degrees.

RETRIEVE

4 Solving problems involving trigonometry

Finding a side of a right-angled triangle in real-life situations

1. A flagpole is 5 m tall.

   A rope from the top of the flagpole meets the ground at an angle of 60°.

   Calculate the length of the rope, to the nearest centimetre.

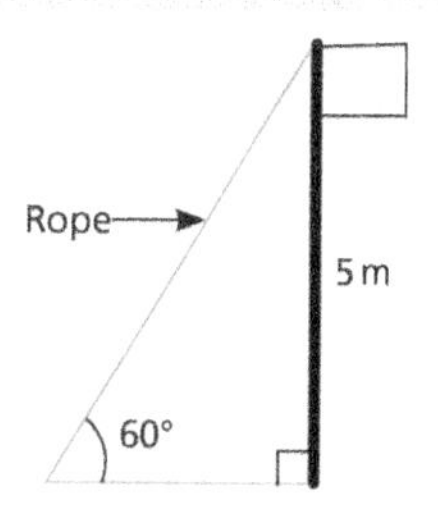

2. A 3-metre ladder leans against a wall.
   The angle between the ladder and the wall is 25°.

   How far up the wall does the ladder reach?
   Give your answer to the nearest centimetre.

Recognising right-angled triangles in other shapes

3. Here is a rectangle.

   Find the length of x.

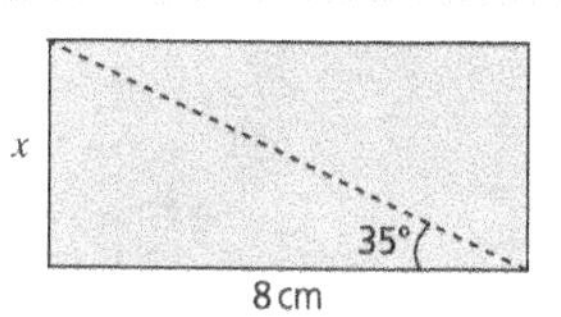

$x =$ ....

4. Here is a rectangle with length 12 cm and width 7 cm.

   Find the size of θ.

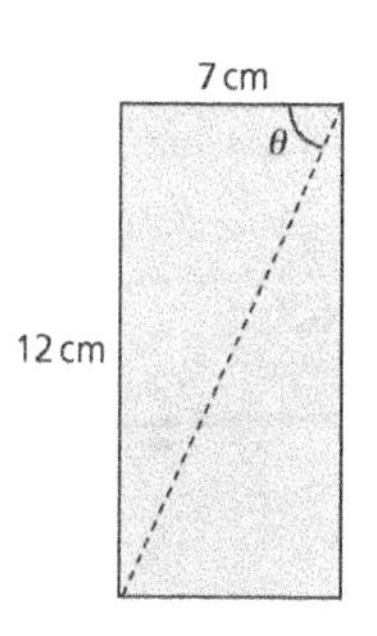

$\theta =$ ....

5. Find the height, h, of each triangle.

   a)

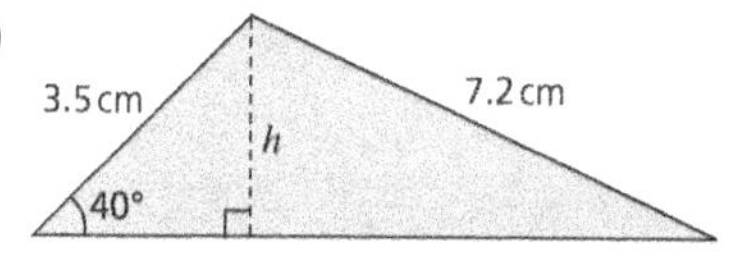

$h =$ ....

   b) Triangle ABC is equilateral

A

6 cm

h

C B

$h =$ ....

ORGANISE 4

Solving problems involving right-angled triangles

Finding a side or an angle of a right-angled triangle in real-life situations

When you know two sides and need a side, use **Pythagoras' theorem**.

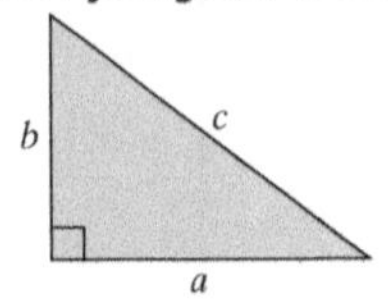

$c^2 = a^2 + b^2$

When you know an angle or need an angle, use **trigonometry**.

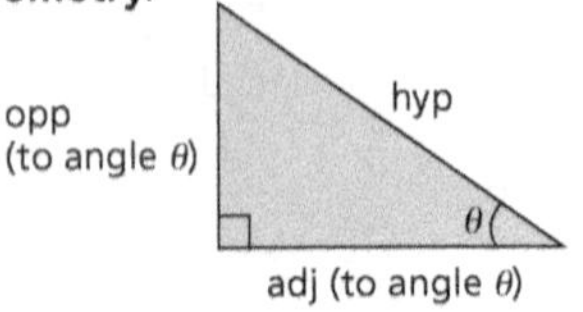

The string of a kite is fastened to the ground at point A.

The kite is directly above point B, 10 m from A.

The string of the kite is 20 m long.

20 m
θ
A
B
10 m

a) Find the angle between the string and the ground, θ.

When you need an angle, use trigonometry.

$\cos = \frac{\text{adj}}{\text{hyp}}$ You know the adj and the hyp, so use the cos ratio.

$\cos\theta = \frac{10}{20}$ For help with using cosine to find angles, see page 40.

$\theta = \cos^{-1}\left(\frac{10}{20}\right) = 60°$

b) Find the height of the kite above point B.

When you know two sides and need a side, use Pythagoras' theorem: $c^2 = a^2 + b^2$. For help with Pythagoras' theorem, see pages 10 and 12.

$20^2 = 10^2 + b^2$ *b* is the unknown height.

$400 - 100 = b^2$

$b = \sqrt{300} = 17.3205\ldots$

$b = 17.32$ m (to the nearest cm)

Finding a side or an angle in shapes

ABC is an isosceles triangle.

AB = AC = 9 cm

CB = 8 cm

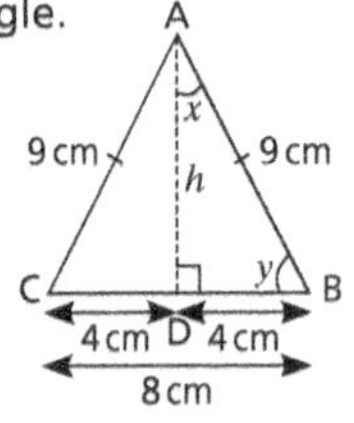

a) Find height h.

In triangle ABD, you know two sides and need a side – use Pythagoras' theorem.

$c^2 = a^2 + b^2$

$9^2 = 4^2 + h^2$

$81 - 16 = h^2$

$h = \sqrt{65} = 8.0622\ldots$

$h = 8.1$ cm (1 d.p.)

b) Find the size of angle x.

To find an angle, use the side lengths you were given in the question – in case you made a mistake calculating h.

A
x
9 cm
hyp
adj
C
B
D
4 cm
opp

To find x:

$\sin x = \frac{\text{opp}}{\text{hyp}}$ You know the opp and hyp, so use the sin ratio.

$\sin x = \frac{4}{9}$

$x = \sin^{-1}\left(\frac{4}{9}\right)$

$x = 26.3877\ldots°$

$x = 26.4°$ (1 d.p.)

c) Find the size of angle y.

Use a fact about angles in a triangle to find y.

$y + 26.4° + 90° = 180°$ Angles in a triangle add up to 180°.

$y = 180 - 26.4° - 90° = 63.6°$

Solving problems involving right-angled triangles

Finding a side or an angle of a right-angled triangle in real-life situations

1. A ramp is made from a 2.7 m length of wood, at an angle of 20° to the ground.

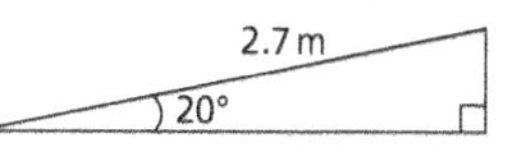

Work out the height of the top of the ramp.

2. Max is building a zip wire from the top of a post to the ground.
The post is 4 m tall.
The wire is 30 m long.

Post
4 m
30 m

Calculate:

a) the horizontal distance from the base of the post to where the wire meets the ground

b) the angle between the wire and the post.

Finding a side or an angle in shapes

3. Here is a rectangle.

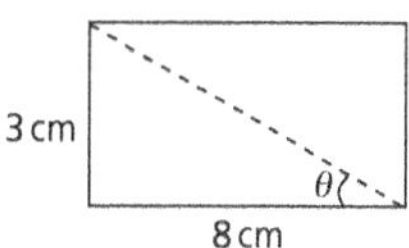

a) Work out angle θ.

b) Find the length of the diagonal of the rectangle.

4. Find the height, x, of this trapezium.

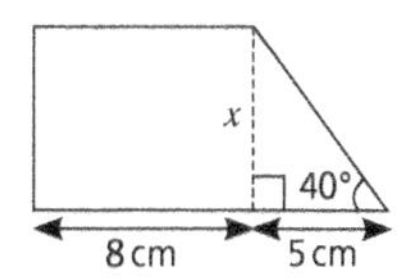

5. ABD and DCB are two right-angled triangles.

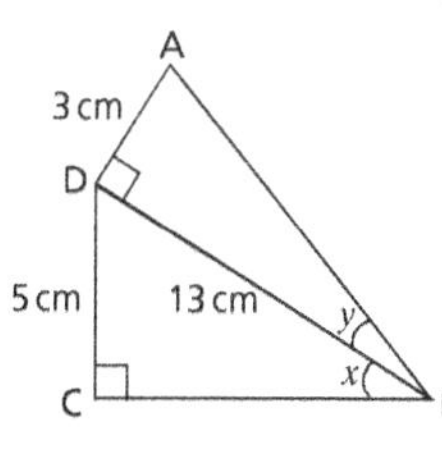

a) Work out x.

b) Work out y.

c) Find the length of AB.

d) Find the length of CB.

ORGANISE

5 Properties of 3D shapes 1

Definitions

3D shapes are **solid shapes or objects** that have three dimensions: length, width and height.

3D means three-dimensional.

Some real-life examples of 3D shapes are a can, a football and a cereal box.

3D shapes have **faces**, **edges** and **vertices**.

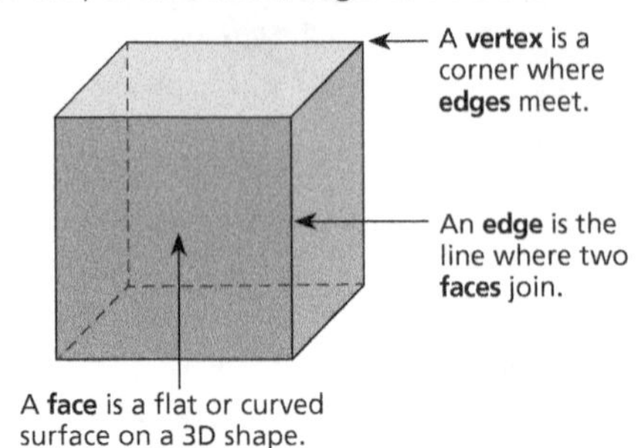

'Vertices' is the plural of 'vertex'.

Properties of 3D shapes

3D shapes with flat faces only

	Cube	Cuboid	Triangular prism	Square-based pyramid
Shape				
Faces	6	6	5	5
Edges	12	12	9	8
Vertices	8	8	6	5

3D shapes with curved faces

	Cylinder	Cone	Sphere
Shape			
Faces	3	2	1
Edges	2	1	0
Vertices	0	1	0

RETRIEVE

5 Properties of 3D shapes 1

Definitions

1. Define what a face of a 3D shape is.

2. Define what an edge of a 3D shape is.

3. Define what a vertex of a 3D shape is.

Properties of 3D shapes

4. Name a 3D shape that has 6 faces, 12 edges and 8 vertices.

5. Name the 3D shape that has 2 faces, 1 edge and 1 vertex.

6. Name the 3D shape that has 5 faces, 8 edges and 5 vertices.

7. Name the 3D shape that has 3 faces, 2 edges and 0 vertices.

8. Name the 3D shape that has 5 faces, 9 edges and 6 vertices.

Properties of 3D shapes 2

Surface area and volume

To work out the total **surface area** of a 3D shape, follow these three steps:

1. Identify the shape of each of the faces of the 3D shape.
2. Work out the area of each of the faces of the 3D shape.
3. Add together the areas of all faces of the 3D shape.

Volume = area of cross section × length

A cross section is the 2D shape that is made when cutting through a 3D shape.

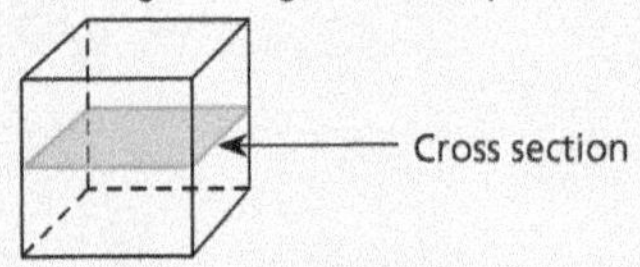

Here is a cube.

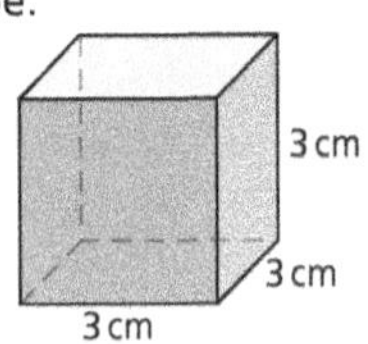

a) **Work out the surface area of the cube.**

A cube has six faces, all of which are squares.

Area of one face: $3 \times 3 = 9\,\text{cm}^2$

Area of six faces: $(3 \times 3) \times 6 = 54\,\text{cm}^2$

Total surface area $= 54\,\text{cm}^2$

b) **Work out the volume of the cube.**

Volume = area of cross section × length

$3 \times 3 \times 3 = 27\,\text{cm}^3$

Here is a cylinder.

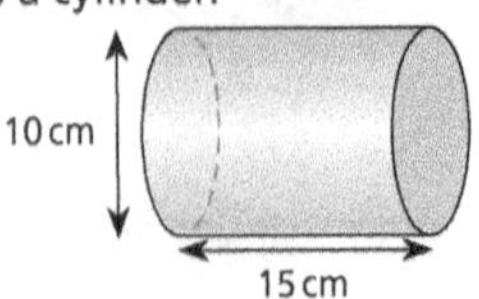

a) **Work out the surface area of the cylinder.**

A cylinder has three faces: two identical circles and one rectangular face. The length of the rectangle is the circumference of either of the circles.

Area of one circle:

$= \pi \times \text{radius}^2$

The radius is half the length of the diameter.

$= \pi \times 5^2$

$= 25\pi\,\text{cm}^2$

Area of two circles: $25\pi \times 2 = 50\pi\,\text{cm}^2$

Area of rectangle =

circle circumference × width

circumference of circle $= \pi \times \text{diameter}$

$= \pi \times 10 = 10\pi\,\text{cm}$

So the area of the rectangle:

$10\pi \times 15 = 150\pi\,\text{cm}^2$

Total surface area:

$50\pi\,\text{cm}^2 + 150\pi\,\text{cm}^2 = 200\pi\,\text{cm}^2$

b) **Work out the volume of the cylinder.**

Volume $= \pi r^2 \times \text{length}$

$= \pi \times 5^2 \times 15$

$= \pi \times 25 \times 15$

$= 375\pi\,\text{cm}^3$

$= 1178.1\,\text{cm}^3$

Compound 3D shapes

A compound 3D shape is made up of two or more 3D shapes.

A solid is formed by placing half a cylinder on a rectangular prism. The solid has a width of 2m, a total height of 3m and a length of 8m.

Work out the volume of the solid. Give your answer to 3 significant figures.

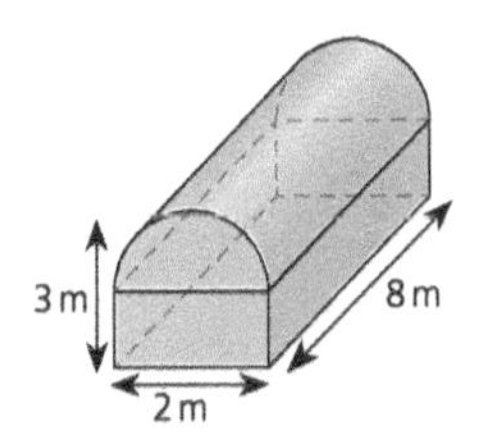

Radius = 1 m

Volume of half cylinder $= \frac{\pi r^2}{2} \times l$

$= \frac{\pi(1)^2}{2} \times 8$

$= 4\pi\,\text{m}^3$

Volume of prism $= 2 \times 2 \times 8 = 32\,\text{m}^3$

Total volume $= 4\pi + 32$

$= 44.6\,\text{m}^3$ (3 s.f.)

5 Properties of 3D shapes 2

Surface area and volume

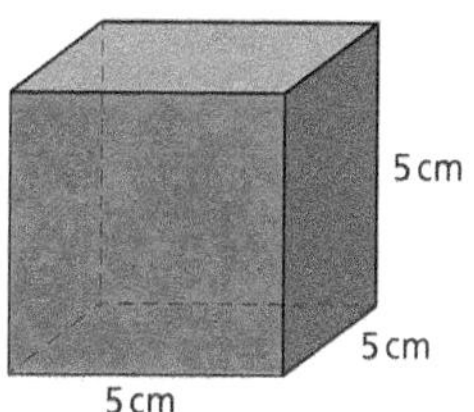

a) Work out the surface area of the cube.

.... cm^2

b) Work out the volume of the cube.

.... cm^3

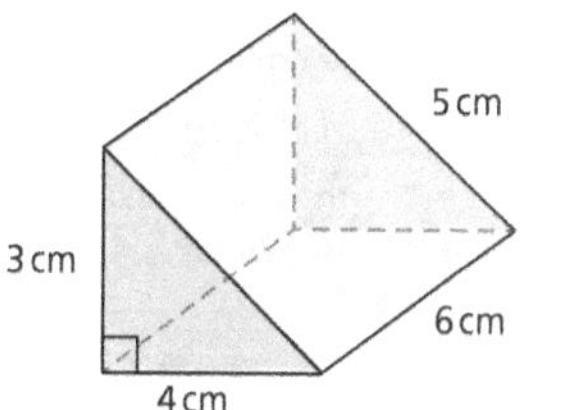

a) Work out the surface area of the triangular prism.

.... cm^2

b) Work out the volume of the triangular prism.

.... cm^3

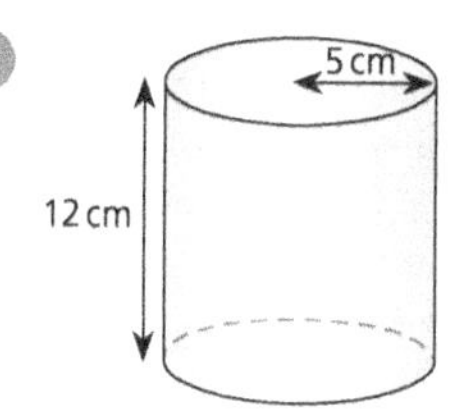

a) Work out the surface area of the cylinder.

.... cm^2

b) Work out the volume of the cylinder.

.... cm^3

Compound 3D shapes

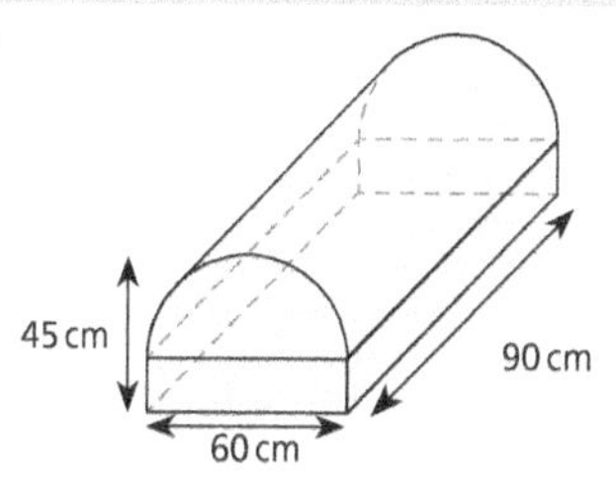

Work out the volume of this compound 3D shape.
Give your answer to 2 significant figures.

.... cm^3

5 Solving problems in 3D

Surface area and volume

The surface area of a 3D shape is the total area of all its faces.

To work out the volume of a shape: **Volume = area of cross section × length**

Each of the 12 cubes below has a surface area of 24cm^2.

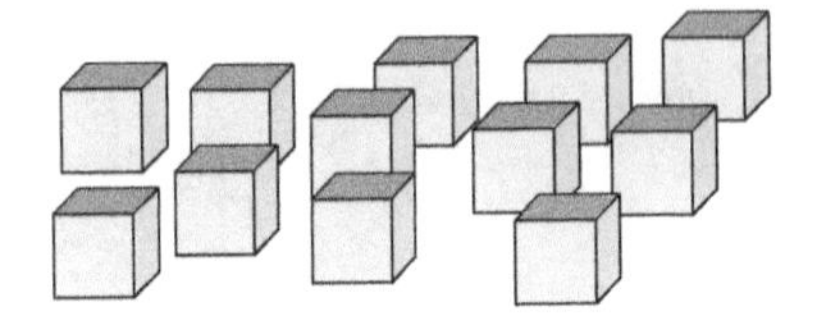

The 12 cubes are joined together to make the cuboid below.

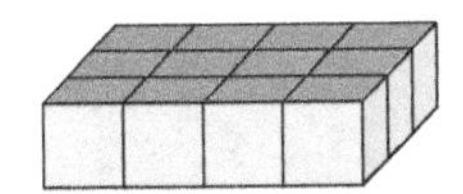

a) What is the surface area of the cuboid?

Each cube has a surface area of 24cm^2. There are six faces on a cube.

Surface area of one face: $24 \div 6 = 4\text{cm}^2$

So side length of the cube is $\sqrt{4} = 2\text{cm}$

Total surface area of the cuboid is:

$((8 \times 2) \times 2) + ((6 \times 2) \times 2) + ((8 \times 6) \times 2)$

$32 + 24 + 96 = 152\text{cm}^2$

Total surface area $= 152\text{cm}^2$

b) What is the volume of this cuboid?

$8 \times 6 \times 2 = 96\text{cm}^3$

Volume $= 96\text{cm}^3$

The length of one cube is 2 cm.

Compound 3D shapes

A compound 3D shape is made up of two or more solids.

When solving a problem, use values that have not been rounded. Only round values at the end.

A component is manufactured by cutting a cylindrical hole of diameter 5 cm through a metal block, as shown.

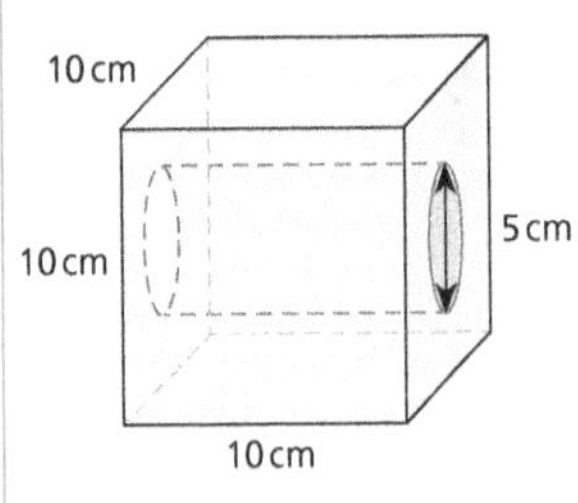

a) Calculate the volume of the block **after** the hole has been drilled.

The shape of the hole is a cylinder.

Volume of the metal block:

$10 \times 10 \times 10 = 1000\text{cm}^3$

Volume of the cylinder:

$(\frac{5}{2})^2 \times \pi \times 10 = \frac{125}{2}\pi\text{cm}^3$

Volume of the block after the cylindrical hole has been drilled is:

$1000 - \frac{125}{2}\pi\text{cm}^3$

$= 803.7\text{cm}^3$ (1 d.p.)

b) What percentage of the block has been removed?

The proportion of the block that has been removed can be expressed as

$\frac{\text{Volume of cylindrical hole}}{\text{Volume of metal block}}$

To convert this proportion to a percentage, multiply by 100:

$\frac{\text{Volume of cylindrical hole}}{\text{Volume of metal block}} \times 100$

$\frac{\frac{125}{2}\pi}{1000} = \frac{1}{16}\pi$

$\frac{1}{16}\pi \times 100 = 19.6\%$

19.6% of the block has been removed.

Solving problems in 3D

Surface area and volume

1 Each of these six cubes has a surface area of $54 cm^2$.

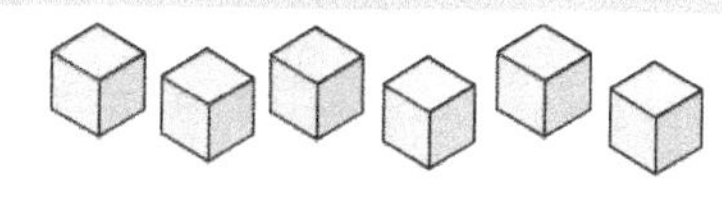

The six cubes are joined together to make the cuboid shown here.

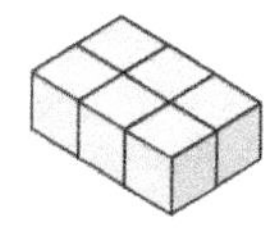

a) Work out the surface area of the cuboid.

.... cm^2

b) Work out the volume of the cuboid.

.... cm^3

2 Each of these six cubes has a surface area of $96 cm^2$.

The cubes are joined together to make this cuboid.

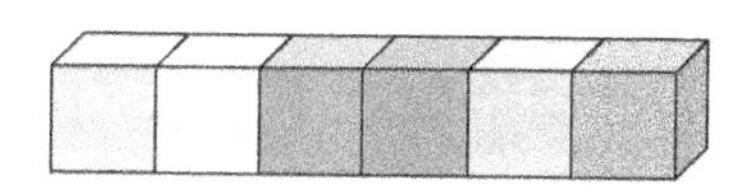

a) Work out the surface area of the cuboid.

.... cm^2

b) Work out the volume of the cuboid.

.... cm^3

Compound 3D shapes

3 A component is manufactured by cutting a cylindrical hole of diameter 8 cm through a metal block, as shown.

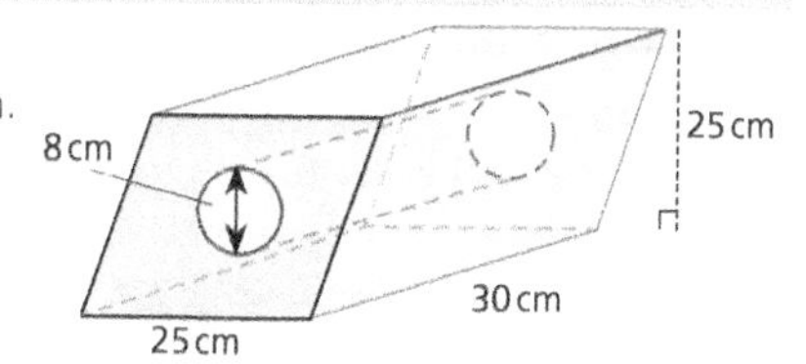

a) Calculate the volume of the block **after** the hole has been drilled.

.... cm^3

b) What percentage of the block has been removed?

.... %

ORGANISE

Introducing standard form

Standard form and powers of 10

Standard form (or **standard index form**) is a way of writing very large or very small ordinary numbers as **powers of 10** so that they are easier to work with.

Recall powers of 10:

$10^0 = 1$

$10^{-1} = \frac{1}{10} = 0.1$ $\qquad 10^1 = 10$

$10^{-2} = \frac{1}{10^2} = 0.01$ $\qquad 10^2 = 10 \times 10 = 100$

$10^{-3} = \frac{1}{10^3} = 0.001$ $\qquad 10^3 = 10 \times 10 \times 10 = 1000$

... and so on.

A number written in standard form is in the form $A \times 10^n$ where $1 \leqslant A < 10$ and n is a positive or negative integer (or zero). The value of A must be equal to or greater than 1 and less than 10.

If the ordinary number is:

- between 0 and 1, n will be negative
- between 1 and 10, n will be zero
- over 10, n will be positive.

A number written to a negative power of 10 in standard form is a decimal number in ordinary form.

Tick the numbers that are correctly written in standard form. Cross the numbers that are not.

a) 9.85×10^4 ✔

The value of A is 9.85 and the value of n is 4.

b) 6.46×10^{-6} ✔

The value of A is 6.46 and the value of n is –6.

c) 37.73×10^2 ✘

The value of A is **not between** 1 and 10.

Converting numbers between ordinary form and standard form

To convert a number **from ordinary form to standard form**, follow these steps:

- Rewrite the ordinary number with the first non-zero digit in the ones place.
- To find n (the power of 10), look at how many places the digits have moved.
- Decide if the value of n is positive, negative or zero.

Write each number in standard form.

a) 7901

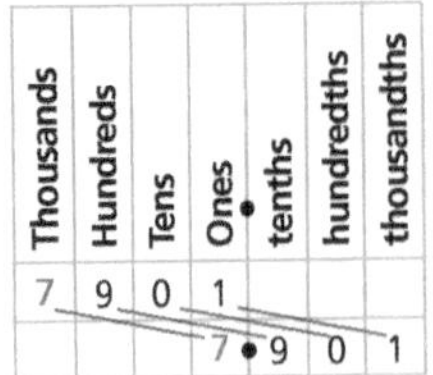

Thousands	Hundreds	Tens	Ones	tenths	hundredths	thousandths
7	9	0	1			
			7 •	9	0	1

The digits have moved by three places and 7901 is greater than 10, so 7901 in standard form is 7.901×10^3

b) 0.027

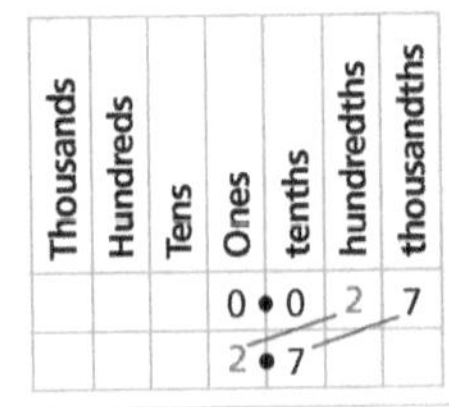

Thousands	Hundreds	Tens	Ones	tenths	hundredths	thousandths
			0 •	0	2	7
			2 •	7		

The digits have moved by two places and 0.027 is between 0 and 1, so 0.027 in standard form is 2.7×10^{-2}

To convert a number **from standard form to ordinary form**, multiply the value of A by the power of 10. Remember to fill in any empty place values with zeros.

Write each number in ordinary form.

a) 5.13×10^4

5.13×10^4 means $5.13 \times 10 \times 10 \times 10 \times 10$, so each digit moves to the left by four places.

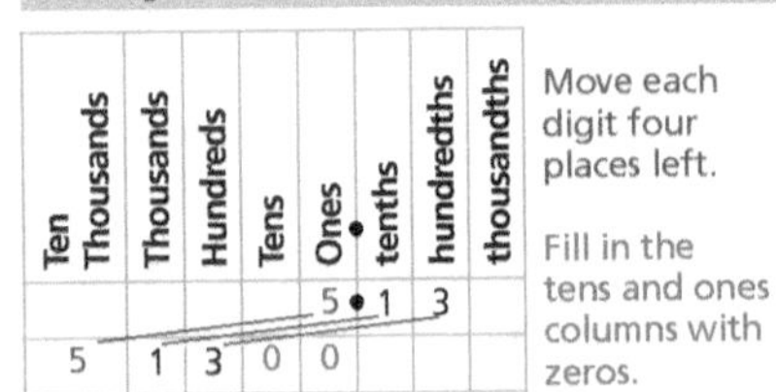

Ten Thousands	Thousands	Hundreds	Tens	Ones	tenths	hundredths	thousandths
				5 •	1	3	
5	1	3	0	0			

Move each digit four places left.

Fill in the tens and ones columns with zeros.

5.13×10^4 is 51 300 as an ordinary number.

b) 1.058×10^{-2}

1.058×10^{-2} means $1.058 \div 10^2$, so each digit moves to the right by two places.

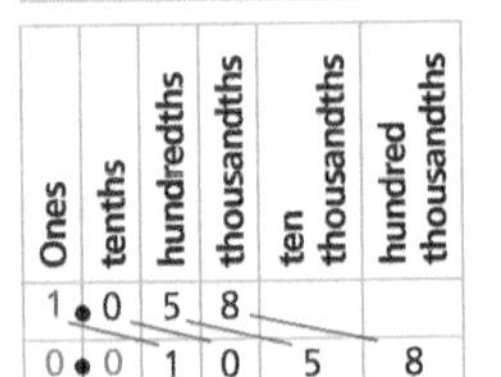

Ones	tenths	hundredths	thousandths	ten thousandths	hundred thousandths
1 •	0	5	8		
0 •	0	1	0	5	8

Move each digit two places right.

Fill in the ones and tenths columns with zeros.

1.058×10^{-2} is 0.01058 as an ordinary number.

RETRIEVE

6 Introducing standard form

Standard form and powers of 10

1 Write each power of 10 as an ordinary number.

a) 10^8 ....

b) 10^{-4} ....

c) 10^6 ....

d) 10^{-8} ....

2 Circle the numbers that are written in standard form.

123.4 $\quad$ 1.234×10^{-10} $\quad$ 12 340 000 $\quad$ 12.34×10^{-2}

12×10^{34} $\quad$ 1.2×10^{34} $\quad$ 1×10^{23} $\quad$ 4×10^0

Converting numbers between ordinary form and standard form

3 Write each ordinary number in standard form.

a) 5326 ....

b) 72 835 000 ....

c) 0.00726 ....

d) 0.209 ....

4 The following numbers are in standard form. Write them as ordinary numbers.

a) 1.12×10^8 ....

b) 6.18×10^{-4} ....

c) 4.19×10^6 ....

d) 2.05×10^{-3} ....

ORGANISE

Comparing and ordering numbers in standard form

Comparing when n is positive

Comparing numbers in standard form is similar to comparing decimals:

- Compare the power of 10 (n).
- The greater number will have the greater power of 10.
- If numbers have the same power of 10, compare the first digit (the highest place value), then the second, and so on.

For example:

2×10^5 is greater than 2×10^3 because 5 is greater than 3.

2.3×10^8 is greater than 8.4×10^7 because 8 is greater than 7.

Which is greater, 5.86×10^6 or 5.79×10^6?

Both numbers have a power of 6. — Look at the powers.

Both numbers have 5 in the ones place. — Look at the highest place value.

5.86×10^6 has an 8 in the tenths place and 5.79×10^6 has a 7 in the tenths place. — Look at the next place value.

So, using inequality notation:

$5.86 \times 10^6 > 5.79 \times 10^6$

Comparing when n is negative

Comparing numbers in standard form when n is negative is the same as comparing when n is positive. Remember that the more negative a number is, the smaller it is.

For example:

3×10^{-2} is greater than 3×10^{-4} because −2 is greater than −4.

6.4×10^{-5} is less than 6.4×10^{-3} because −5 is less than −3.

3.8×10^{-8} is greater than 8.3×10^{-12} because −8 is greater than −12.

7×10^{-3} is less than 7×10^2 because −3 is less than 2.

3.46×10^2 is greater than 3.46×10^{-1} because 2 is greater than −1.

Which is greater, 2.33×10^{-5} or 4.33×10^{-5}?

Both numbers have a power of −5.

Comparing the ones places, 2 is less than 4.

So, using inequality notation:

$2.33 \times 10^{-5} < 4.33 \times 10^{-5}$

Ordering numbers in standard form

To order numbers in standard form, identify the greatest and least numbers first, then order from there using the same procedure for comparing numbers in standard form.

Order these numbers from smallest to greatest:

3.1×10^7 $\quad$ 9.2×10^7 $\quad$ 5.85×10^{-5} $\quad$ 5.42×10^{-4} $\quad$ 3.14×10^{10}

Identify the smallest and greatest numbers in the list.

5.85×10^{-5} is the smallest number because −5 is the smallest power.
3.14×10^{10} is the greatest number because 10 is the greatest power.

Then look at the powers of the remaining numbers and order them.

5.42×10^{-4} has a power of −4.
Both 3.1×10^7 and 9.2×10^7 have a power of 7, but 3.1 is less than 9.2, so 3.1×10^7 is less than 9.2×10^7

5.85×10^{-5} $\quad$ 5.42×10^{-4} $\quad$ 3.1×10^7 $\quad$ 9.2×10^7 $\quad$ 3.14×10^{10}

Smallest ⟵⟶ Largest

So, using inequality notation: $5.85 \times 10^{-5} < 5.42 \times 10^{-4} < 3.1 \times 10^7 < 9.2 \times 10^7 < 3.14 \times 10^{10}$

RETRIEVE

Comparing and ordering numbers in standard form

Comparing when n is positive

1 Compare each pair of numbers in standard form. Write your answer using inequality notation.

a) 1.35×10^8 and 1.38×10^{10}

....

b) 6.95×10^5 and 6.9×10^5

....

c) 9×10^2 and 9×10^3

....

d) 6.239×10^6 and 1.53×10^5

....

Comparing when n is negative

2 Compare each pair of numbers in standard form. Write your answer using inequality notation.

a) 4.09×10^8 and 4.09×10^{-10}

....

b) 9.64×10^{-5} and 9.6×10^{-5}

....

c) 1.8×10^{-5} and 1.8×10^{-2}

....

d) 4.21×10^{-6} and 2.592×10^2

....

Ordering numbers in standard form

3 Order this set of numbers from **greatest to least**. Write your answer using inequality notation.

8.06×10^6 $\quad$ 2.64×10^0 $\quad$ 7.8×10^6 $\quad$ 2.6×10^{-9} $\quad$ 3.03×10^{-7}

....

7 Distance–time graphs

Return journeys

A distance–time graph shows time on the x-axis and distance travelled on the y-axis.

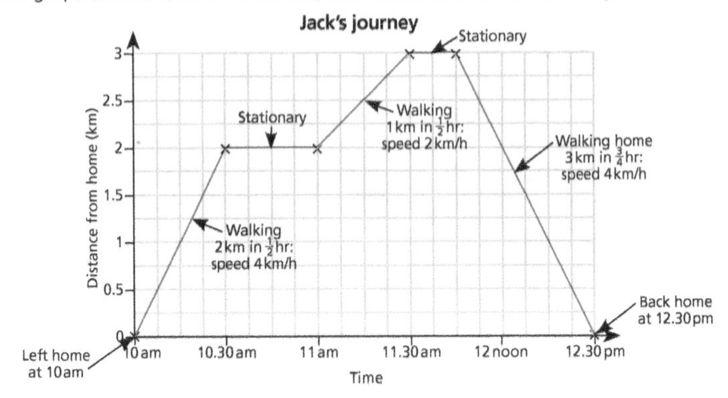

Make sure you understand what the gradient of the line on a distance–time graph represents:

- A **positive gradient** (/) shows that an object/person is **moving away from the starting point**.
- A **horizontal line** means that an object/person has **stopped** moving.
- A **negative gradient** (\) shows that an object/person is **moving back towards the starting point**.

One-way journeys

The steeper the line, the faster the speed of travel.

Here is the distance–time graph for two runners in a race.

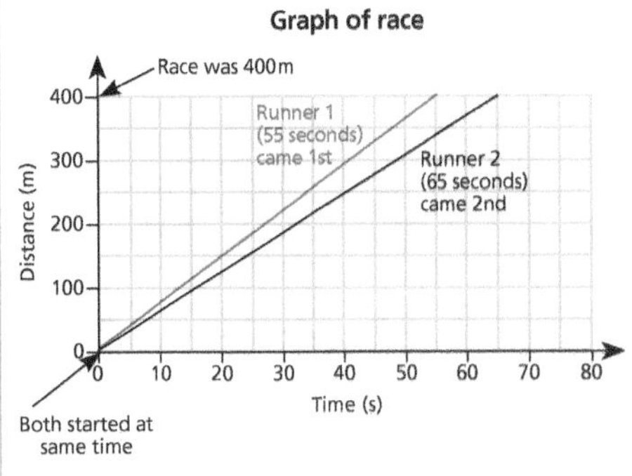

Here is the distance–time graph for two trains from Aytown to Beetown.

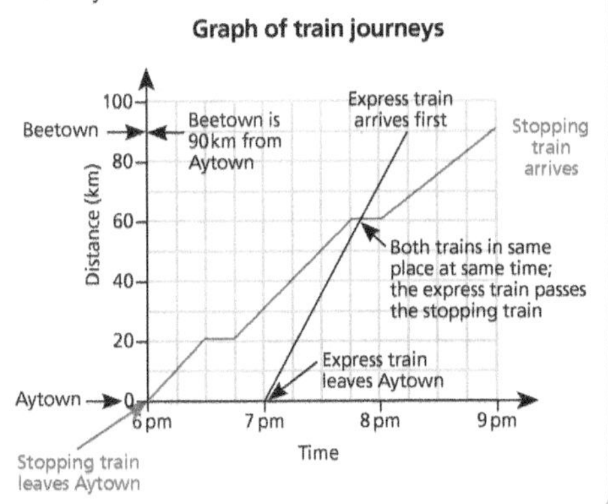

Distance–time graphs

Return journeys

1. The graph shows Jill's cycle ride from home to her friend's house, and back.

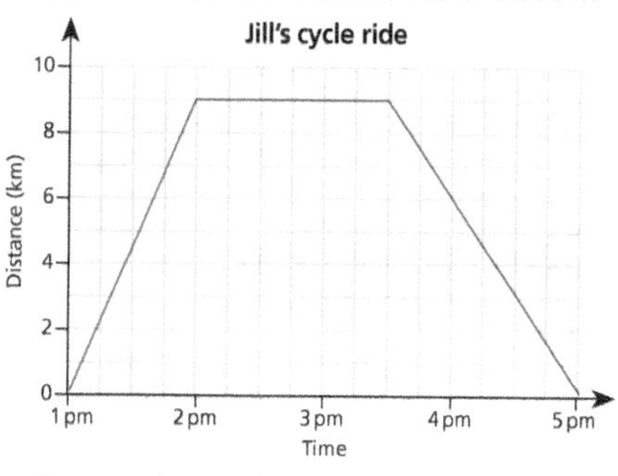

a) What is the distance from Jill's house to her friend's house?

b) How long does Jill stay at her friend's house?

c) At what speed does Jill cycle on her way to her friend's house?

d) How long is Jill away from home?

e) When did Jill cycle fastest: on the way to her friend's or on the way home?

Explain how you know.

f) How far did Jill cycle in total?

One-way journeys

2. Amit leaves home at 9 am to drive to York.
He drives 70 km in 1 hour. Then he stops for a break for 30 minutes.
He then drives for another 120 km and arrives in York at 12 o'clock.

a) Draw a distance–time graph to represent his journey.

Remember to give your graph a title.

b) When was Amit driving faster: before or after his break?

c) How far is it from Amit's home to York?

3. The graph shows two bus journeys 100 km long.

a) Which bus set off first?

b) How much later did the second bus set off?

c) How far had they each travelled when Bus B passed Bus A?

d) Which bus arrived first at the destination?

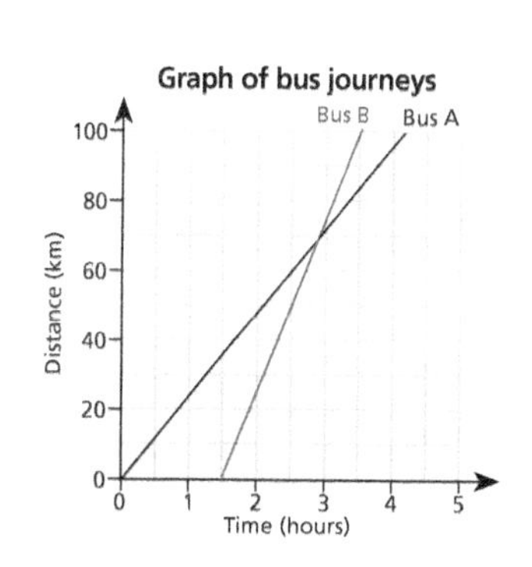

ORGANISE

7 Interpreting other real-life graphs

Velocity–time graphs

Velocity–time graphs show how acceleration changes over time. Time is on the horizontal (x) axis and velocity (or speed) is on the vertical (y) axis. Make sure you understand what the gradient of the line on a velocity–time graph represents.

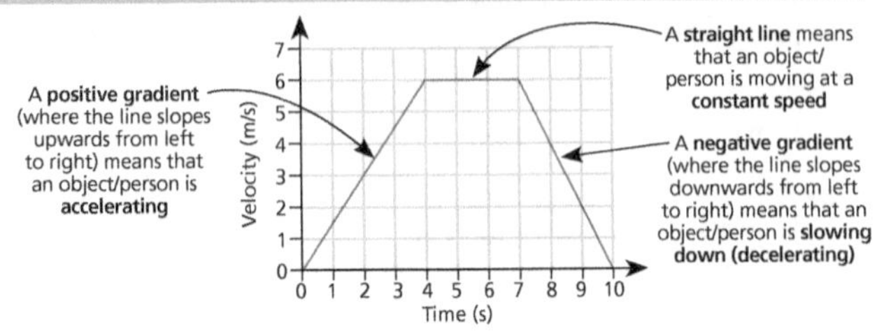

The speed of a car is recorded at 20-second intervals and shown on this speed–time graph. Describe the car's speed.

Between 0 and 20 seconds, the car is accelerating.

Between 20 and 40 seconds, the car is accelerating but at a slower rate because the gradient has decreased.

Between 40 and 60 seconds, the car is moving at a constant speed.

Between 60 and 80 seconds, the car is slowing down (decelerating) until it stops moving at 80 seconds.

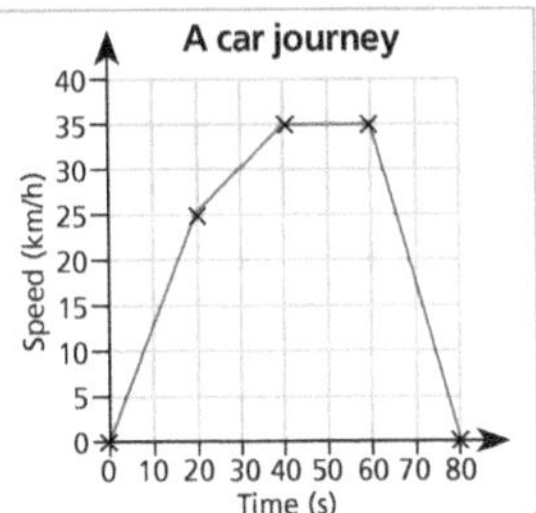

Conversion graphs

Conversion graphs are straight line graphs that can be used to convert from one unit to another. They are often used for currency conversions and measurement conversions.

This graph converts between pounds (£) and euros (€) using an exchange rate on a particular day. How many pounds would you get for €100?

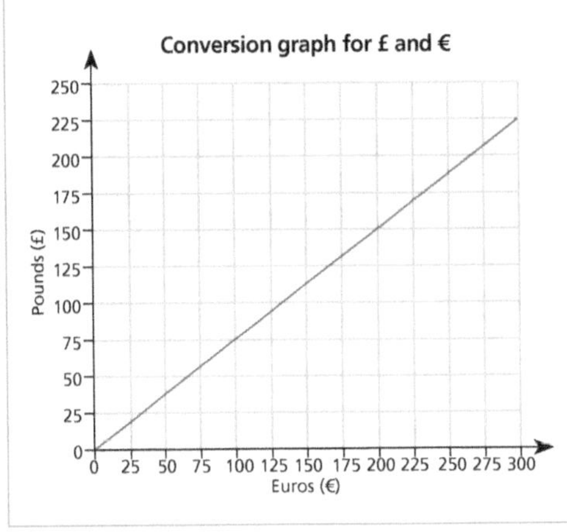

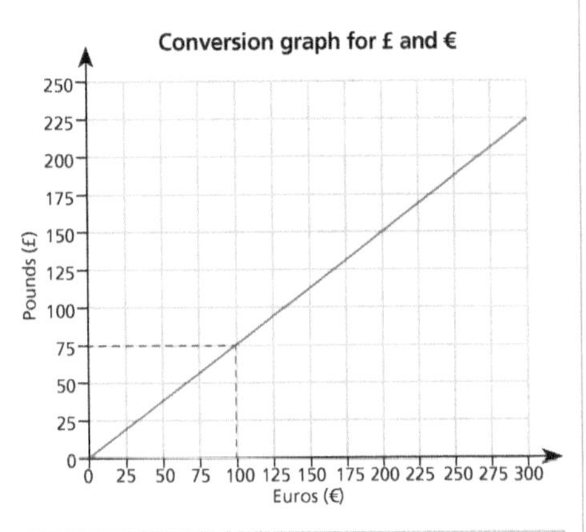

When interpreting real-life graphs, make sure you consider:

- the gradient of the line
- the y-intercept.

Use a ruler to draw a vertical line from the amount in the currency you have been given, until you reach the graph. Then draw a horizontal line to the other axis to find how much it converts to.

According to the graph, €100 = £75

7 Interpreting other real-life graphs

Velocity–time graphs

1. The velocity–time graph shows a 50-second car journey.

Describe the car's journey.

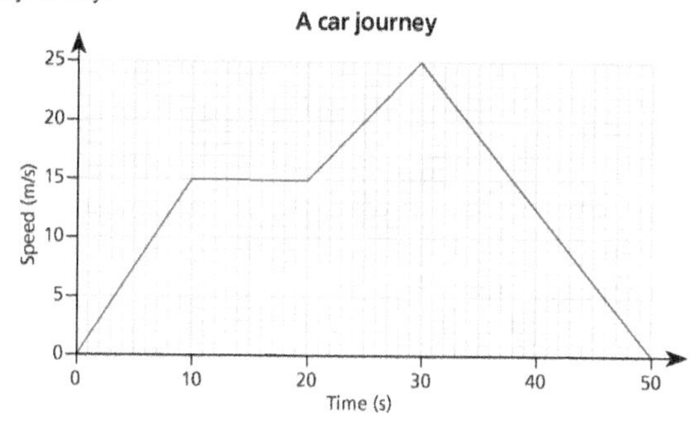

Conversion graphs

2. The graph converts between pounds (£) and US dollars ($) using an exchange rate on a particular day.

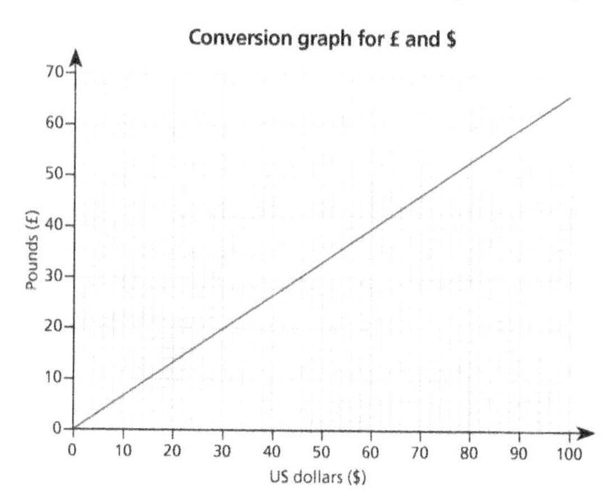

a) How much is $60 worth in pounds? £

b) How much is £20 worth in US dollars? $

7 Plotting real-life graphs

Velocity–time graphs

A car accelerates for the first 2 seconds that it moves. It then travels at a constant speed of 30 m/s for 3 seconds. The car then accelerates again for the next 3 seconds, reaching a speed of 50 m/s.

Draw a velocity–time graph to represent this information.

'A car accelerates for the first 2 seconds that it moves': draw a line sloping upwards from 0 to 2 seconds on the time axis

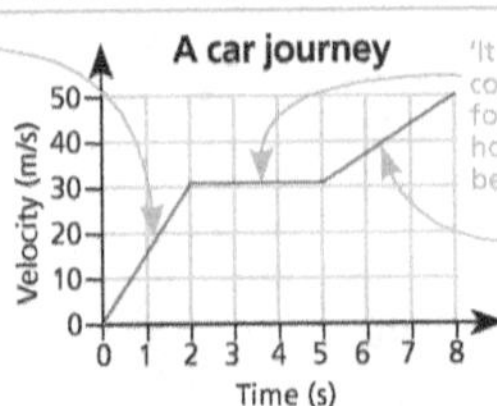

'It then travels at a constant speed of 30 m/s for 3 seconds': draw a horizontal line at 30 m/s between 2 and 5 seconds

'The car then accelerates again for the next 3 seconds, reaching a speed of 50 m/s': draw a line sloping upwards from 5 to 8 seconds

Conversion graphs

When there is a table of values, a conversion graph is plotted in the same way as a linear graph in the form $y = mx + c$. The first row of the table of values is usually plotted against the x-axis and the second row against the y-axis.

Use the information in the table to draw a conversion graph for inches and centimetres.

Inches	0	2	6	10
Centimetres	0	5	15	25

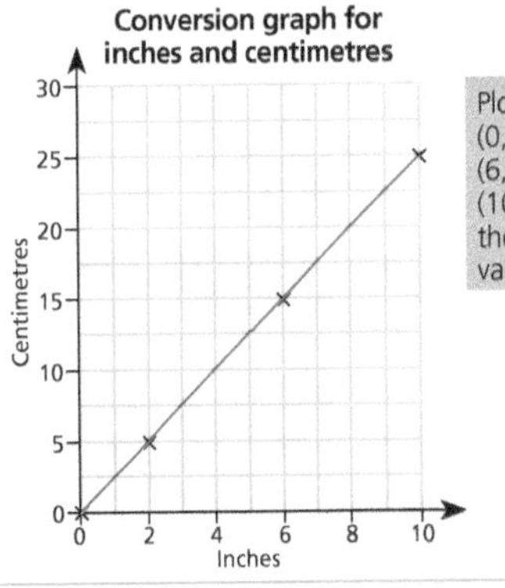

Plot the points (0, 0), (2, 5), (6, 15) and (10, 25) from the table of values.

An electrician charges a £20 call-out fee plus £30 per hour of work she does.

a) Complete the table of costs for different lengths of jobs.

Time (hours)	0	1	2	3	4	5	6
Cost (£)							

The charge at 0 hours is £20 because this is the call-out fee. For every hour, the cost increases by £30.

Time (hours)	0	1	2	3	4	5	6
Cost (£)	20	50	80	110	140	170	200

b) Plot a graph of cost against time.

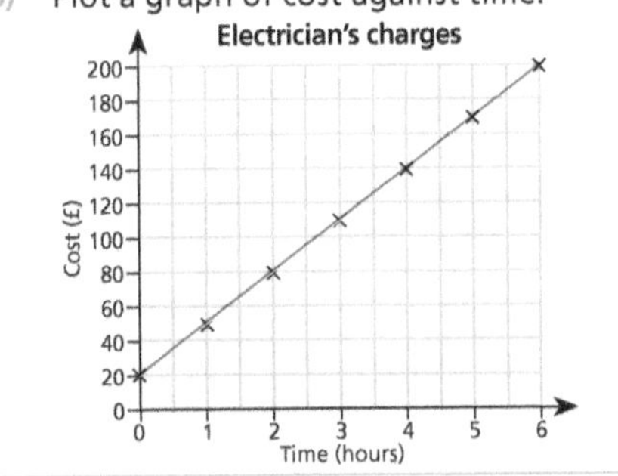

A conversion is given that 5 miles = 8 km.

a) Complete the table using this fact.

Miles	5	10	20	25	50
Kilometres					

The two units are in direct proportion. This means that if one part of the ratio is multiplied or divided by an amount, the same is done to the other part, e.g.:

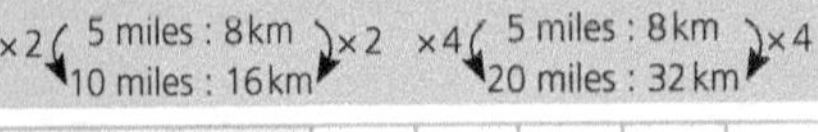

Miles	5	10	20	25	50
Kilometres	8	16	32	40	80

b) Draw a conversion graph from the point (0, 0) to represent this information.

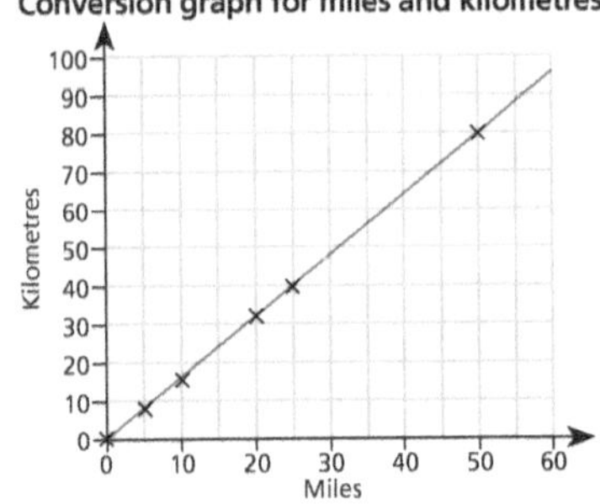

Plotting real-life graphs

Velocity–time graphs

1. A car accelerates for the first 2 seconds that it moves. It then travels at a constant speed of 40 m/s for 8 seconds. The car then slows down for 1 second, and then travels at a constant speed of 20 m/s for 2 seconds.

   Draw a velocity–time graph to represent this information.

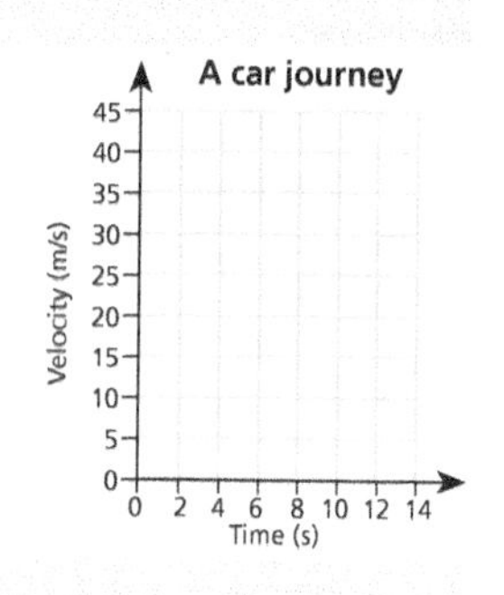

Conversion graphs

2. A taxi driver charges customers a fixed amount of £3 plus an extra 50p for every mile travelled.

   a) Use this information to complete the table.

Distance (miles)	0	1	2	4	6	8	10
Total cost (£)							

   b) Use this information to complete the graph showing the total cost to customers for journeys of up to 10 miles.

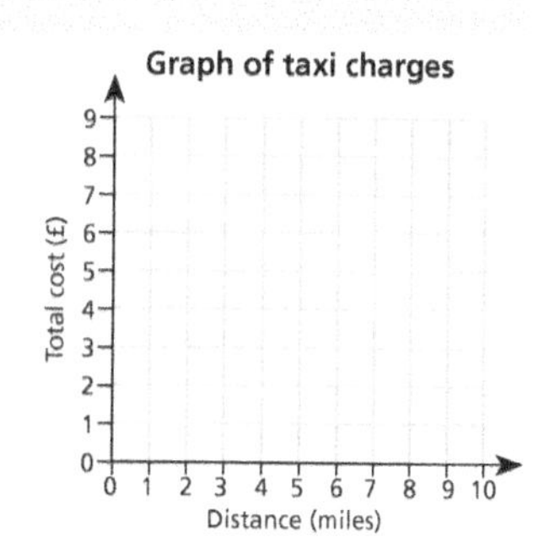

3. **a)** A conversion is given that £1 = €1.20

   Complete the table using this fact.

Pounds (£)	1.00	10.00		25.00	
Euros (€)	1.20		24.00		60.00

   b) Draw a conversion graph to represent this information. Start by plotting the point (0, 0).

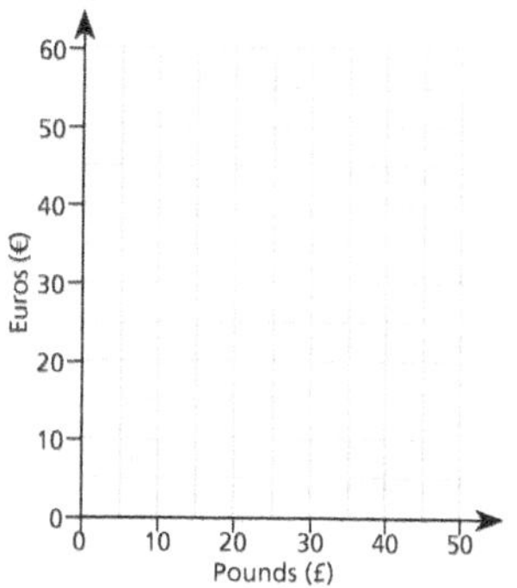

4. **a)** A conversion is given that 2 inches = 5 cm.

   Complete the table using this fact.

Inches	2	4	12	20	25
Centimetres					

   b) Draw a conversion graph to represent this information. Start by plotting the point (0, 0).

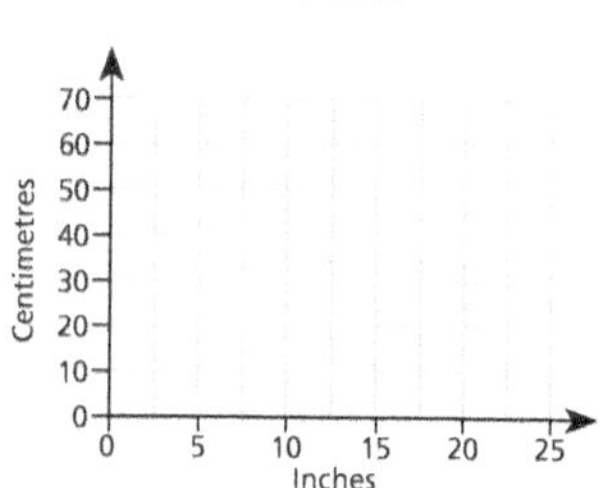

ORGANISE 7

Rearranging linear equations to $y = mx + c$

Rearranging linear equations

Linear equations can be written in the form $y = mx + c$ where m represents the gradient and c represents the y-intercept.

The gradient describes the steepness of a line, and whether the line slopes upwards or downwards. The y-intercept is the point where the graph crosses the y-axis.

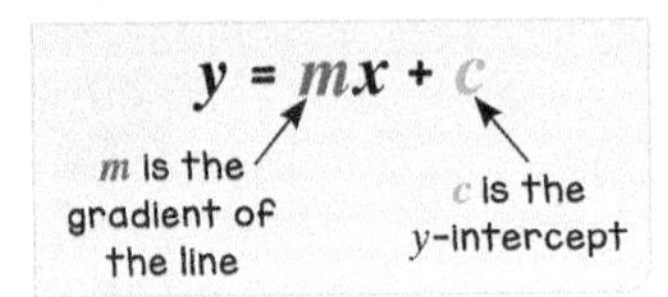

Write down the gradient and the y-intercept of the straight line $y = 3x + 1$

Gradient = 3 y-intercept = 1

To rearrange linear equations into the form $y = mx + c$, use inverse operations.

Rearrange the equation $2y - 12x = 14$ into the form $y = mx + c$ and find the gradient and the y-intercept.

$2y - 12x = 14$ (+ 12x both sides)

$2y = 12x + 14$ (÷ 2 both sides)

$y = 6x + 7$

Isolate the variable y so that it is the only term on the left of the equals sign. Use inverse operations to move terms.

Gradient = 6 y-intercept = 7

Rearrange the equation $2y + 4 = 8x$ into the form $y = mx + c$ and find the gradient and the y-intercept.

$2y + 4 = 8x$ (− 4 both sides)

$2y = 8x - 4$ (÷ 2 both sides)

$y = 4x - 2$

Gradient = 4 y-intercept = −2

Rearrange the equation $3y - 6 = 6x + 9$ into the form $y = mx + c$ and find the gradient and the y-intercept.

$3y - 6 = 6x + 9$ (+6 +6) Add 6 to both sides.

$3y = 6x + 15$ (÷ 3 ÷ 3 ÷ 3) Divide both sides by 3.

$y = 2x + 5$

Gradient = 2 y-intercept = 5

When dividing the equation by a number, remember to divide **all** terms by the number.

Equivalent linear equations

To find equivalent linear equations, rearrange the equations so that both are in the form $y = mx + c$.

Which of these equations represents the same straight line as $2x + y = 8$?

$y = 2x + 8$ $y + 8 = 2x$ $y = -2x + 8$

Rearrange the equation $2x + y = 8$

$2x + y = 8$ (− 2x both sides)

$y = -2x + 8$

So the answer is $y = -2x + 8$

RETRIEVE 7

Rearranging linear equations to $y = mx + c$

Rearranging linear equations

1 a) Rearrange the equation $2y - 10x = 6$ into the form $y = mx + c$.

....

b) Find the gradient and the y-intercept.

Gradient = ...

y-intercept = ...

2 a) Rearrange the equation $2y + 8 = 6x$ into the form $y = mx + c$.

....

b) Find the gradient and the y-intercept.

Gradient = ...

y-intercept = ...

3 a) Rearrange the equation $2y + 2 = 4x$ into the form $y = mx + c$.

....

b) Find the gradient and the y-intercept.

Gradient = ...

y-intercept = ...

4 a) Rearrange the equation $4y + 3 = 4x - 5$ into the form $y = mx + c$.

....

b) Find the gradient and the y-intercept.

Gradient = ...

y-intercept = ...

5 a) Rearrange the equation $y - 2 = -4x + 7$ into the form $y = mx + c$.

....

b) Find the gradient and the y-intercept.

Gradient = ...

y-intercept = ...

Equivalent linear equations

6 Which of these equations represents the same straight line as $3x + y = 10$? Circle your answer.

$y = -3x + 10$ $\quad$ $y - 3x = 10$ $\quad$ $y = 3x - 10$

7 Which of these equations represents the same straight line as $4x + 2y = 6$? Circle your answer.

$y = -4x + 6$ $\quad$ $y = -2x + 3$ $\quad$ $y = -2x - 6$

Using graphs to solve problems

Reading values from a graph

To read a graph, look at:

- the title – what is the graph about?
- the axis labels – what does each axis show you?
- the numbers on the axes – what is the scale and what does one square on each axis represent?

This graph shows the cost of hiring a bike.

Find the cost of hiring a bike for:

a) exactly 2 hours

b) $4\frac{1}{4}$ hours

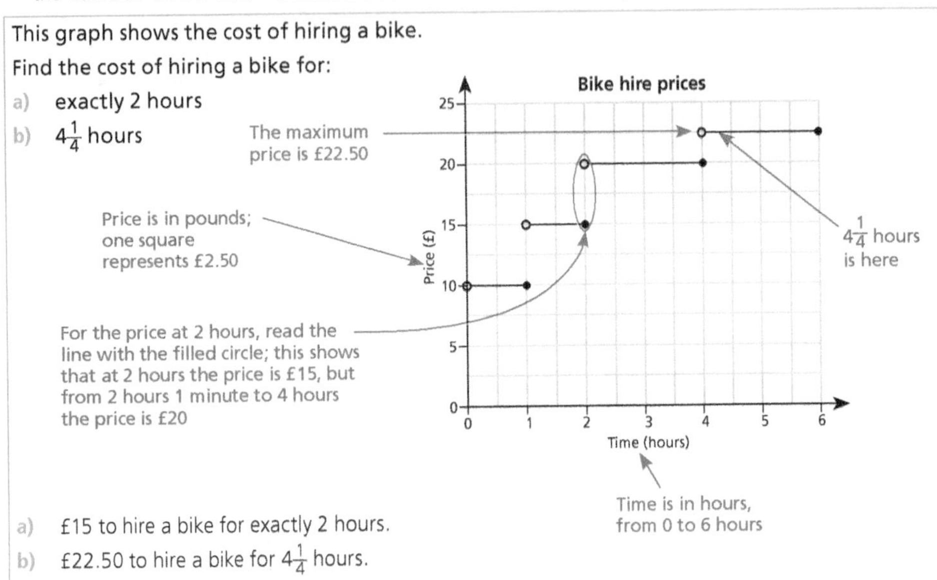

a) £15 to hire a bike for exactly 2 hours.

b) £22.50 to hire a bike for $4\frac{1}{4}$ hours.

Estimating values from a graph

When the graph does not pass exactly through grid points, you need to estimate the value.

The graph shows the temperature of a cup of tea.

Estimate the temperature of the tea after 10 minutes.

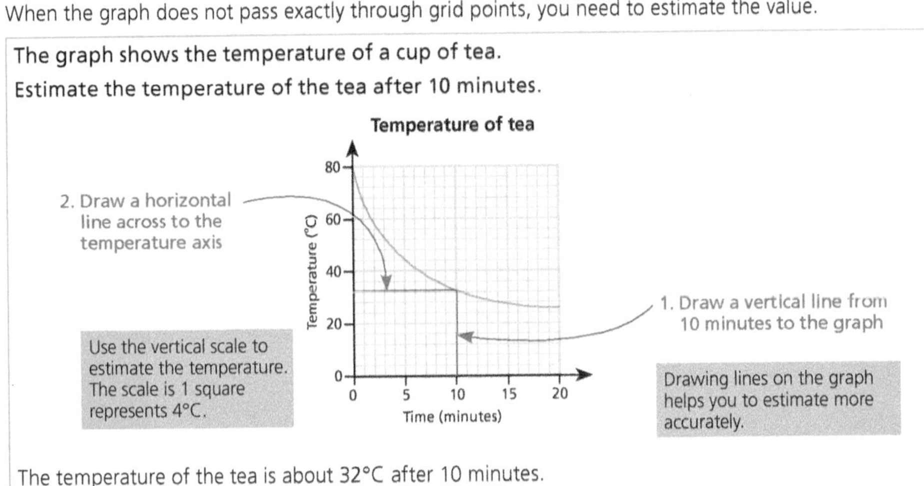

The temperature of the tea is about 32°C after 10 minutes.

RETRIEVE

7 Using graphs to solve problems

Reading values from a graph

1 Here is a graph of postage prices.

a) Use the graph to find the postage price for:

i) a 150g letter

ii) a 700g parcel

b) It costs £4.70 to post a parcel weighing more than 750g and up to and including 1000g.

Draw a line on the graph to show this.

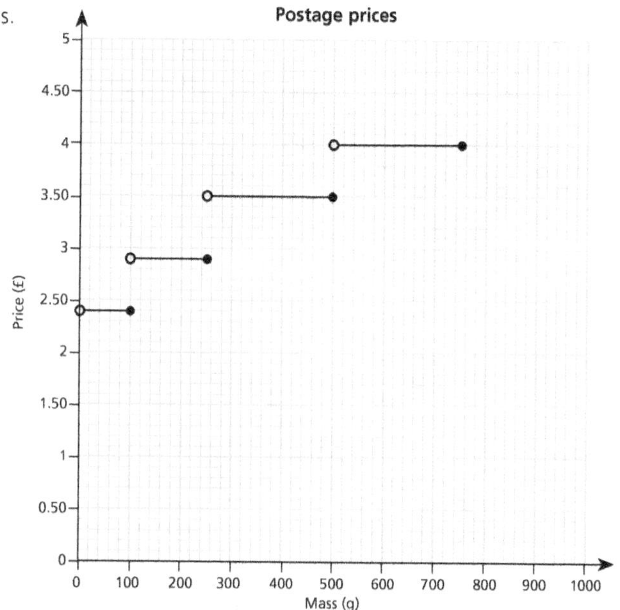

Estimating values from a graph

2 The graph shows the number of bacteria cells in a sample.

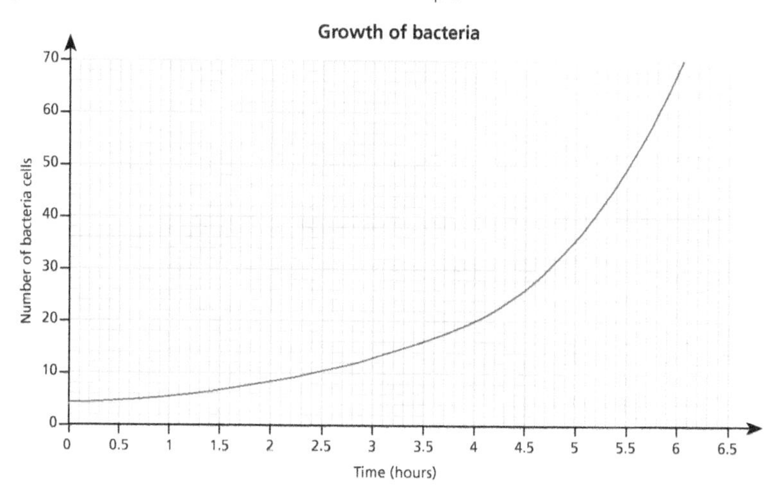

a) Estimate the number of bacteria cells after 5.25 hours.

b) Estimate the time when there are 30 bacteria cells.

ORGANISE 7

Using graphs to solve simultaneous linear equations

Reading solutions from a graph

On a graph, the solution to a pair of simultaneous equations is the point where the graphs cross. The solution is the x-value and y-value at this point.

Two equations, with two unknowns ⟶ Two solution values

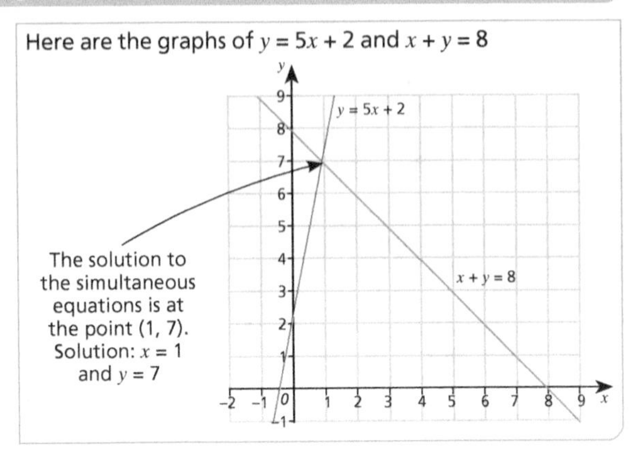

Drawing graphs to solve simultaneous equations

To solve two simultaneous equations graphically, draw their graphs and find the point where they cross.

Solve these simultaneous equations graphically: $y = 3x - 4$
$x + 2y = 6$

To draw the graphs, you can make a table of values for each one, or use $y = mx + c$.

Making tables of values

$y = 3x - 4$

x	−2	−1	0	1	2
y	−10	−7	−4	−1	2

Plot the points and join them with a straight line.

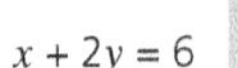

$x + 2y = 6$

For equations in this form, use a table of values with $x = 0$ and $y = 0$

x	0	6
y	3	0

When $y = 0$, $x = 6$
$x = 6$

When $x = 0$, $2y = 6$
$y = 3$

Using $y = mx + c$

$y = 3x - 4$ intercepts the y-axis at (0, −4) and has gradient 3.

Make y the subject of $x + 2y = 6$

See page 36 for changing the subject of a formula.

$x + 2y = 6$

$2y = -x + 6$ Subtract x from both sides.

$y = -\frac{1}{2}x + 3$ Divide both sides by 2.

$x + 2y = 6$ intercepts the y-axis at (0, 3) and has gradient $-\frac{1}{2}$

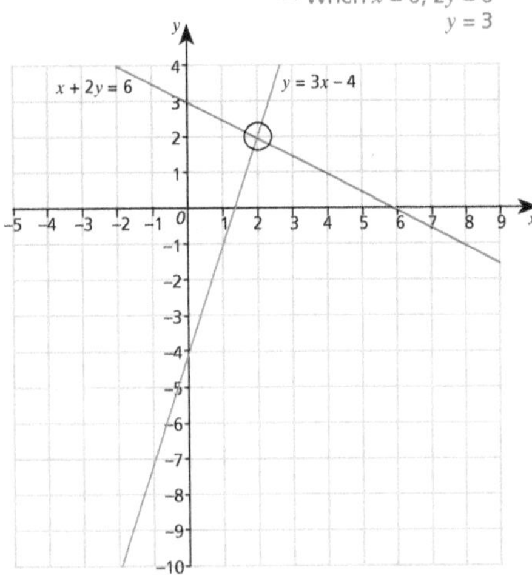

Solution: $x = 2$, $y = 2$

Using graphs to solve simultaneous linear equations

Reading solutions from a graph

1. Here are the graphs of three equations:

$x + y = 5$ $y = 2x + 2$ $y = x + 5$

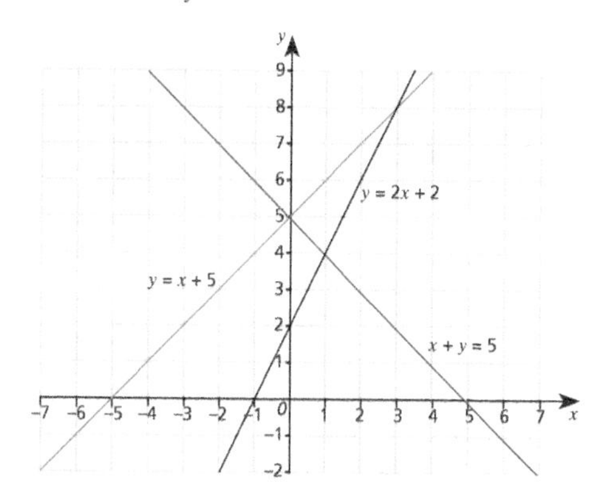

Find the solutions to:

a) $y = x + 5$ and $x + y = 5$ $x =$ ...., $y =$ ....

b) $x + y = 5$ and $y = 2x + 2$ $x =$ ...., $y =$ ....

c) $y = 2x + 2$ and $y = x + 5$ $x =$ ...., $y =$ ....

Drawing graphs to solve simultaneous equations

2. Solve these simultaneous equations graphically:

$y = 2x - 1$

$3x + 2y = 12$

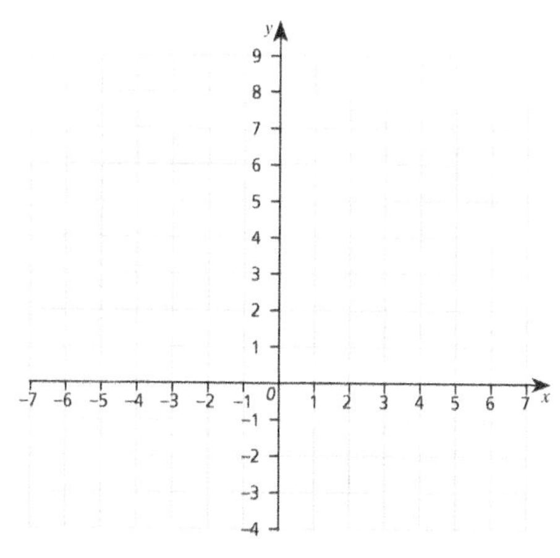

$x =$ ...., $y =$ ....

Graphing inequalities

Lines parallel to the axes

To draw the graph of an inequality, draw the graph line and shade the side of the line where the points satisfy the inequality.

Draw the graphs of the inequalities $x > -3$ and $y \leqslant 4$.

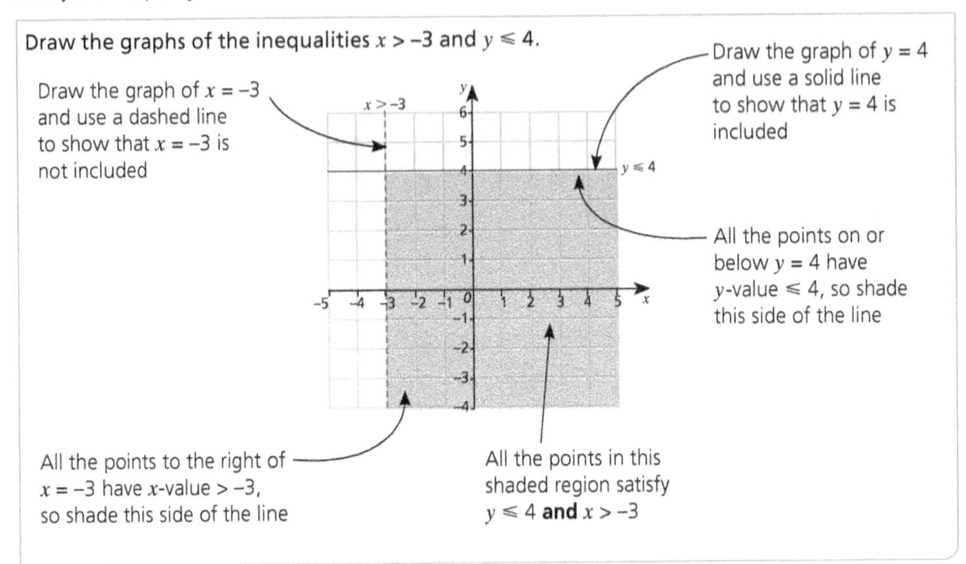

Sloping lines

To decide which side of the line to shade, test points on each side to see if they satisfy the inequality.

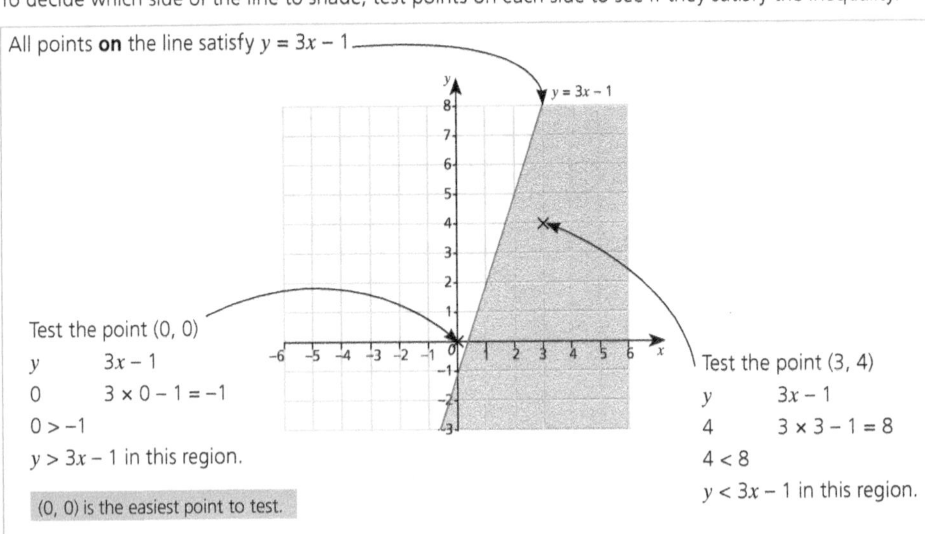

RETRIEVE

7 Graphing inequalities

Lines parallel to the axes

1 **a)** Draw the graphs of the inequalities $x \leqslant 5$ $y > -2$ $x > -5$ $y \leqslant 3$

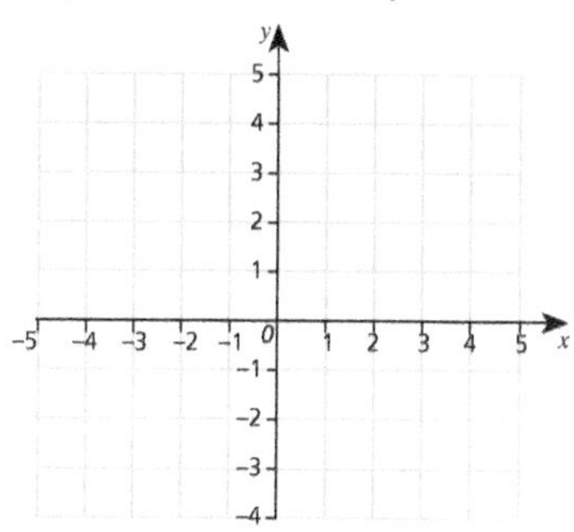

b) Shade the region that satisfies all four inequalities.

Sloping lines

2 Shade the region that satisfies the inequalities $y \leqslant 2x + 1$ $y \geqslant x$ $y \leqslant 6$

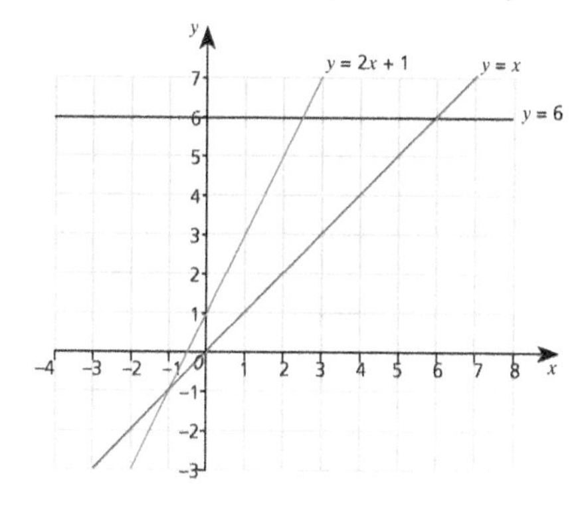

3 Draw the graphs of the inequalities $y < x + 2$ $y \geqslant -2x$

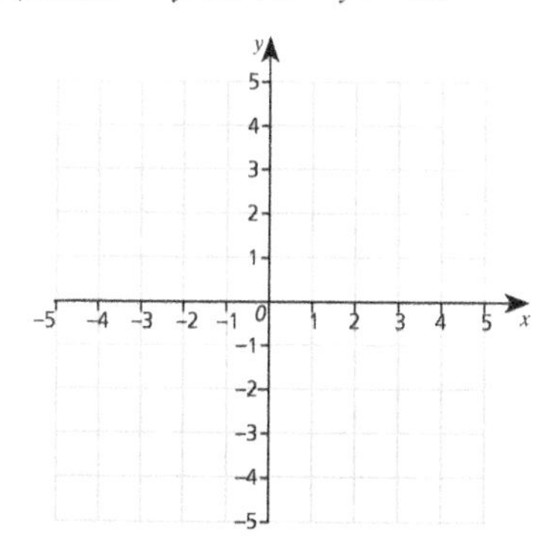

7 Quadratic graphs

Drawing a quadratic graph

The equation of a quadratic graph has an x^2 term and no higher power of x.

Draw the graph of $y = x^2 + 4$ for $-3 \leqslant x \leqslant 3$

1. Make a table of values. Include the x-values given in the question.

$-3 \leqslant x \leqslant 3$

x	−3	−2	−1	0	1	2	3
x^2	9	4	1	0	1	4	9
$y = x^2 + 4$	13	8	5	4	5	8	13

Make a row for x^2

2. Plot the points.
3. Join them with a smooth curve; don't make the bottom of the curve pointed.

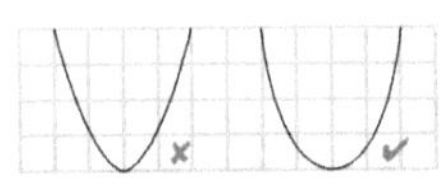

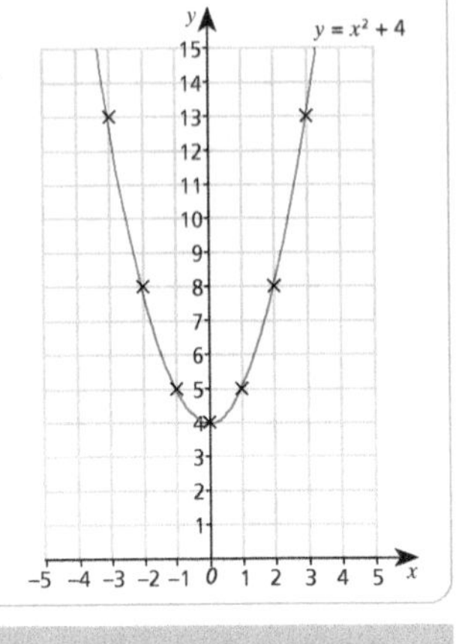

Recognising quadratic graphs

A quadratic graph is a U-shaped curve, called a **parabola**.

Quadratic graph with a positive x^2 term

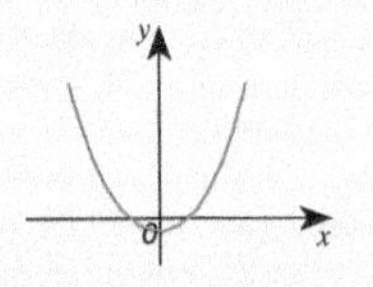

Quadratic graph with a negative x^2 term

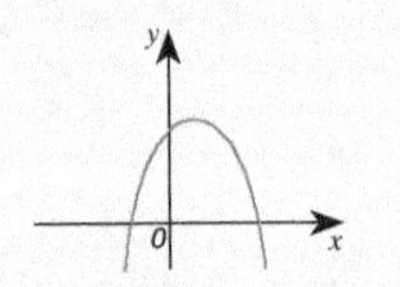

A quadratic graph is symmetrical.

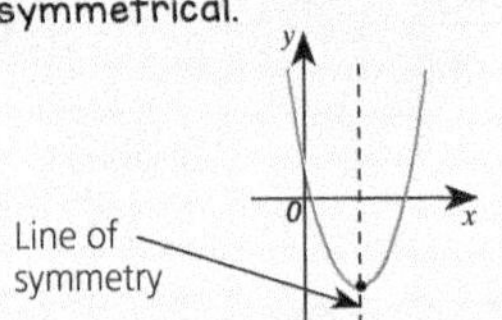

Match each graph below to one of these equations:

$y = x^3$ $y = -x^2 + 2$ $y = x^2 - 3$ $y = x - 3$

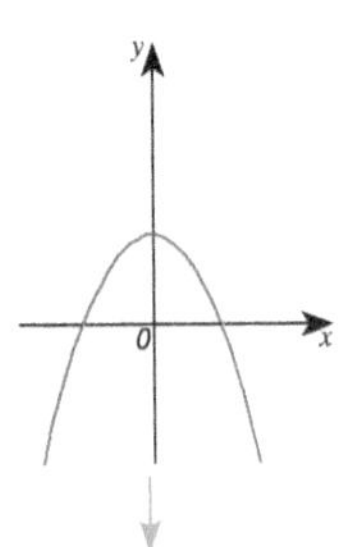

'Upside down' U-shaped curve – the graph is quadratic with a negative x^2 term.

$y = -x^2 + 2$

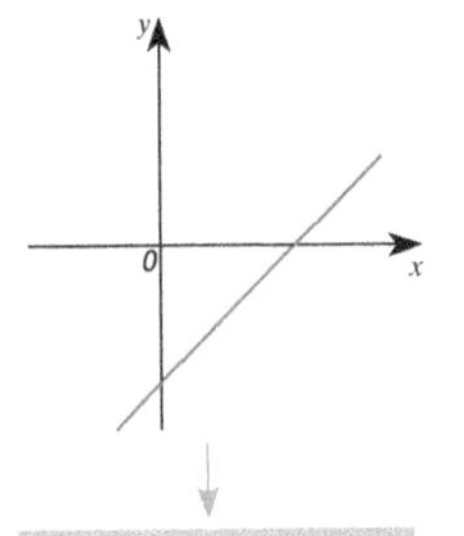

A linear graph has an equation of the form $y = mx + c$

$y = x - 3$

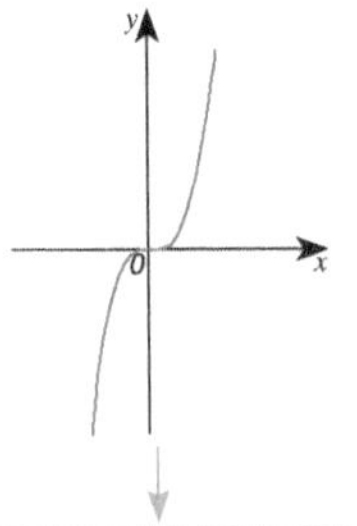

First identify the graphs that you can. This one has the equation that is left over.

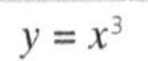
$y = x^3$

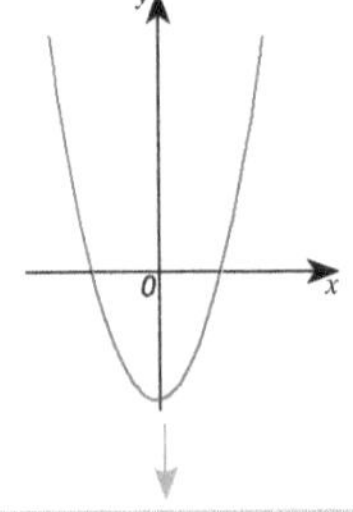

U-shaped curve – the graph is quadratic with a positive x^2 term.

$y = x^2 - 3$

Quadratic graphs

Drawing a quadratic graph

1 Draw the graphs of the following for $-3 \leqslant x \leqslant 3$:

a) $y = -x^2$

x	-3			0			
x^2	9						
$y = -x^2$	-9						

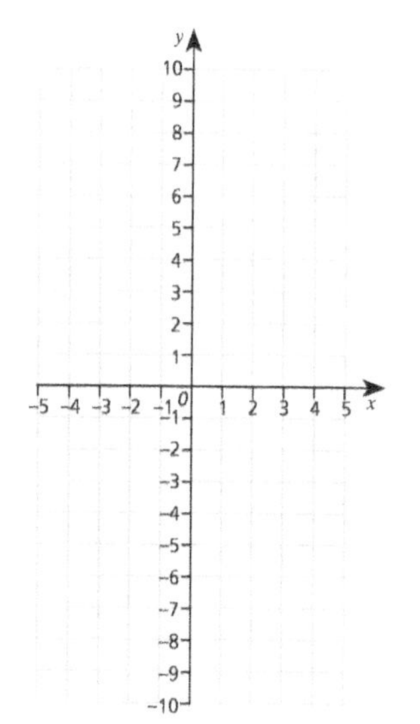

b) $y = x^2 + 1$

x	-3			0			
x^2	9						
$y = x^2 + 1$	10						

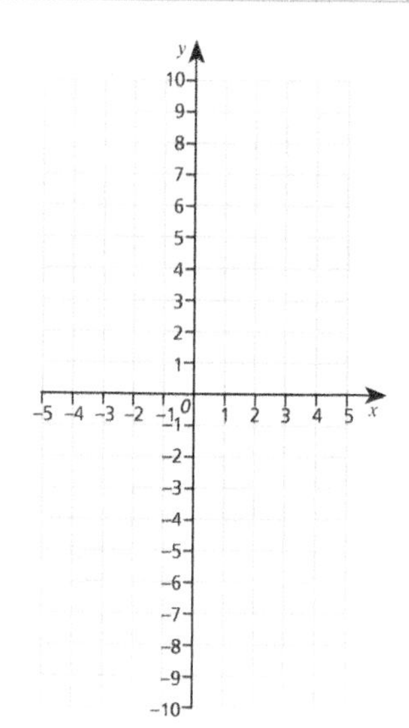

Recognising quadratic graphs

2 Match each graph to its equation.

$y = \frac{1}{x}$	$y = -x^2 + 3x - 1$	$y = -2x - 1$	$y = x^2 + 2$

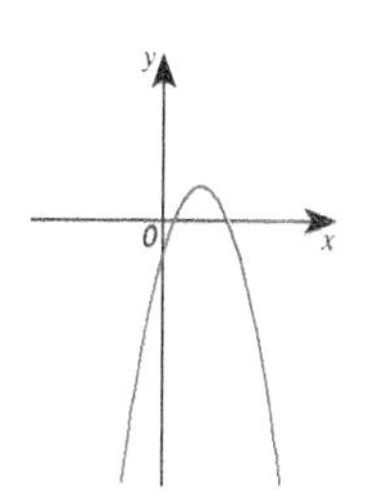

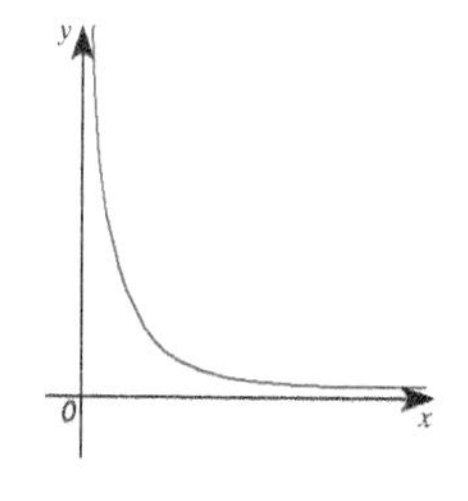

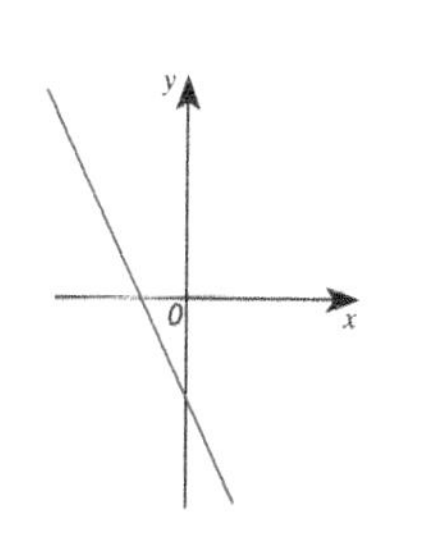

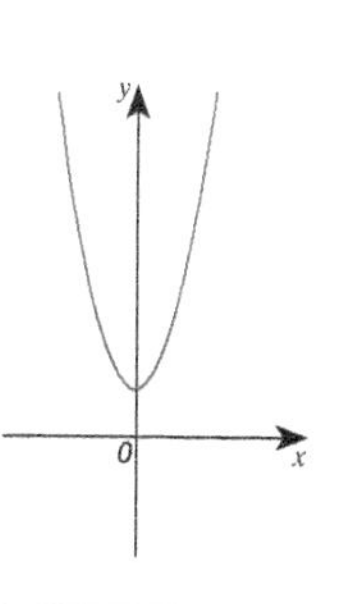

Mixed questions

1 **a)** How many lines of symmetry does this shape have?

b) What is the order of symmetry of this shape?

2 A single card is taken from a standard deck of 52 playing cards.

Calculate the probability of:

a) choosing a red card

b) choosing a face card

c) choosing the ace of hearts

3 **a)** Find the product of $(3x - 1)(x + 2)$

b) Find the product of $(3x - 1)(x + 2)(2x - 5)$

4 Solve each equation.

a) $3(a - 2) = 7a + 5$

$a =$ ....

b) 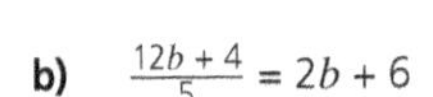

$\frac{12b + 4}{5} = 2b + 6$

$b =$ ....

5 Are these two triangles congruent?

Give a reason to support your answer.

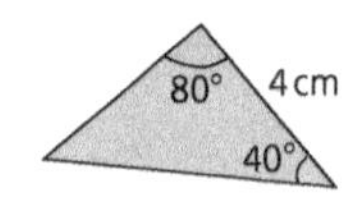

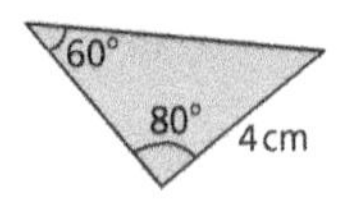

6 **a)** Solve the inequality $3(x - 3) < 15 - x$

b) Given that the solution to **a)** is a positive integer, list all the possible values for x.

7. The line AB is drawn on a 1 cm grid.

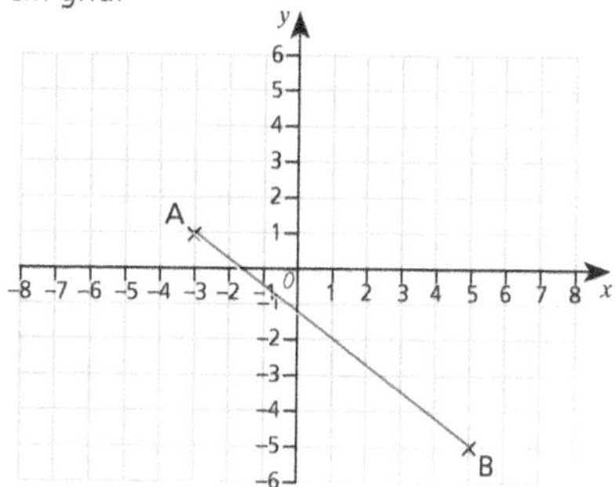

Use Pythagoras' theorem to work out the length of AB.

Draw a right-angled triangle using AB as the diagonal.

.... cm

8. The formula for the volume of a cylinder is given as $V = \pi r^2 h$ where r is the radius and h is the height.

a) Rearrange the formula to make h the subject.

....

b) Find the height of a cylinder with volume 300 cm³ and radius 5 cm.

.... cm

9. Complete the following Venn diagram using the numbers 1–20.

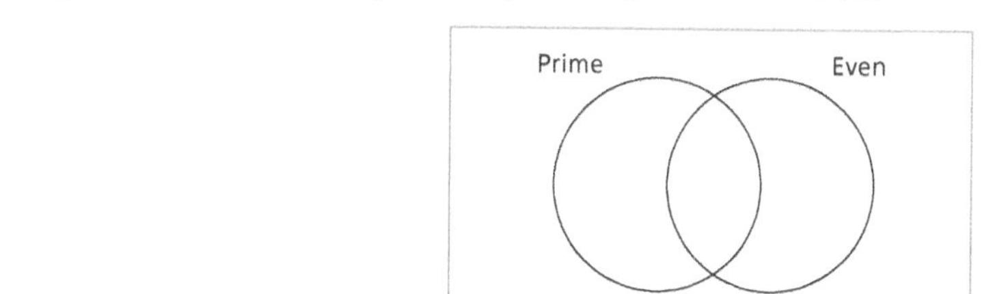

10. Work out the value of x in each of the following triangles.

a)

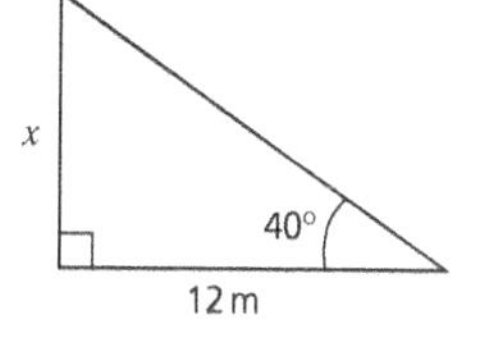

b)

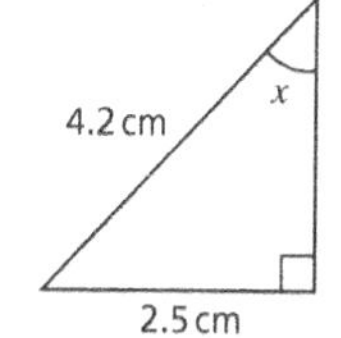

$x =$ .... m

$x =$ .... °

Mixed questions

11 **a)** The average distance between Earth and the Moon is 384 400 km.

Write this distance in standard form.

.... km

b) The average distance between Earth and the Sun is 1.5×10^8 km.

Write this distance in ordinary form.

.... km

12 **a)** Which of the following equations represents a line with the same gradient as $3x - y = 7$?
Circle your answer.

$x + 3y = 5$ $y - 3x = 10$ $y = \frac{1}{3}x + 7$

b) Which of the following equations represents a line with the same y-intercept as $2x + y + 3 = 0$?
Circle your answer.

$y = 5 - 2x$ $4x - y - 3 = 0$ $2y - x = 3$

13 **a)** Find the volume of the box shown below.

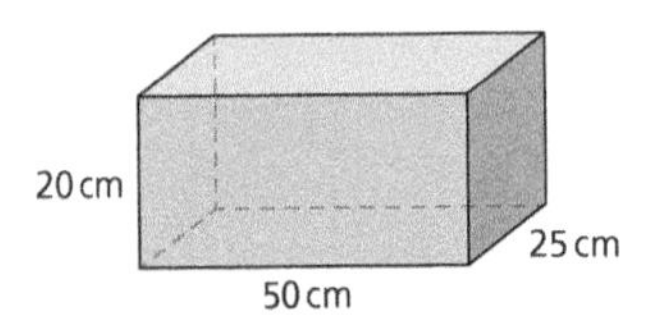

.... cm^3

b) The box is to be used to store smaller packages, as shown below.

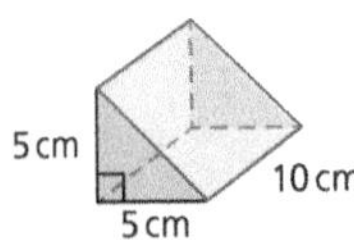

How many of these packages can be stored in the box?

....

14 For safety reasons, a ladder should make an angle of 75° with the ground.
A 3-metre long ladder is propped against a wall.

a) How far away from the base of the wall should the bottom of the ladder be?
Give your answer to the nearest centimetre.

.... cm

b) How far up the wall will the ladder reach?
Give your answer in metres, correct to 1 decimal place.

.... m

Mixed questions

15 Solve these simultaneous equations graphically: $x + y = 5$ $3x - 2y = 0$

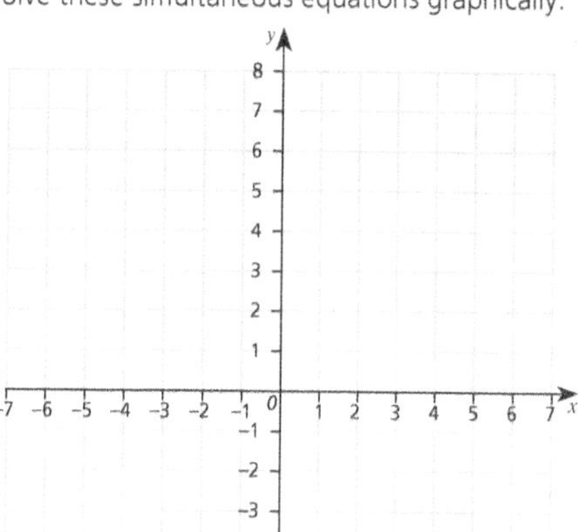

$x =$ ...., $y =$ ....

16 **a)** A cyclist leaves home at 10 am and cycles at a speed of 10 km/h for one hour before stopping for a 15-minute break.

They cycle a further 6 km in 30 minutes before taking another 15-minute break. They then return home, cycling at a steady speed with no further stops, arriving at 1 pm.

Draw a distance–time graph to represent the journey.

b) What was the cyclist's speed on the return journey?

.... km/h

17 **a)** Draw the graphs of the inequalities

$x \geqslant -1$ $y \geqslant x - 2$ $x + y \leqslant 3$

Shade the region that satisfies all three inequalities.

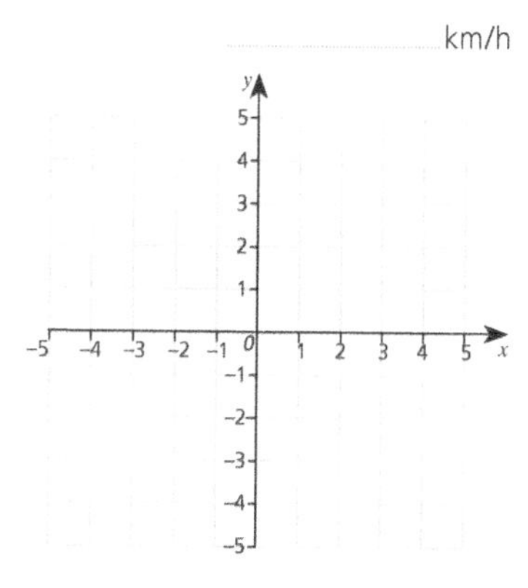

b) Which of the following points satisfies all three inequalities? Circle your answer.

(1, 1) (2, 2) (2, −1)

Key facts and vocabulary

Number

Sequence of square numbers	1, 4, 9, 16, 25, 36, 49, …
Sequence of triangular numbers	1, 3, 6, 10, … (+2, +3, +4, +5)
Standard form	A number written as $A \times 10^n$, where $1 \leqslant A < 10$ and n is an integer Large numbers have positive n: $4.3 \times 10^6 = 4300000$ (6 places) Numbers less than 1 have negative n: $2.5 \times 10^{-4} = 0.00025$ (4 places)

Algebra

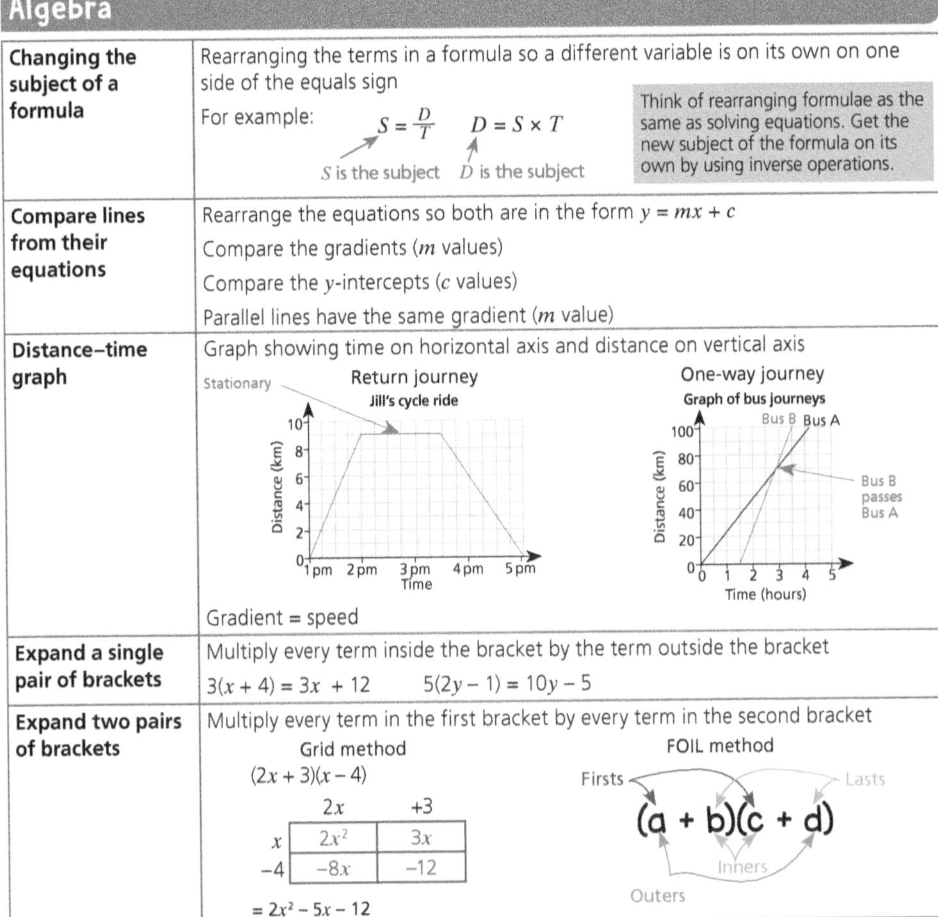

Changing the subject of a formula	Rearranging the terms in a formula so a different variable is on its own on one side of the equals sign For example: $S = \frac{D}{T}$ (S is the subject) $D = S \times T$ (D is the subject) Think of rearranging formulae as the same as solving equations. Get the new subject of the formula on its own by using inverse operations.
Compare lines from their equations	Rearrange the equations so both are in the form $y = mx + c$ Compare the gradients (m values) Compare the y-intercepts (c values) Parallel lines have the same gradient (m value)
Distance–time graph	Graph showing time on horizontal axis and distance on vertical axis Return journey: Jill's cycle ride (Stationary) One-way journey: Graph of bus journeys (Bus B passes Bus A) Gradient = speed
Expand a single pair of brackets	Multiply every term inside the bracket by the term outside the bracket $3(x + 4) = 3x + 12$ $5(2y - 1) = 10y - 5$
Expand two pairs of brackets	Multiply every term in the first bracket by every term in the second bracket Grid method: $(2x + 3)(x - 4)$ x: $2x^2$, $3x$; -4: $-8x$, -12 $= 2x^2 - 5x - 12$ FOIL method: $(a + b)(c + d)$ — Firsts, Outers, Inners, Lasts

Key facts and vocabulary

Expand three pairs of brackets	Expand the first pair of brackets: $(2x + 3)(x - 4)(x - 3) = (2x^2 - 5x - 12)(x - 3)$ Then multiply every term in the first bracket by every term in the second bracket: $(2x^2 - 5x - 12)(x - 3) = 2x^3 - 11x^2 + 3x + 36$
Formula	A rule connecting two or more variables or quantities The formula for speed is: Speed = $\frac{\text{distance}}{\text{time}}$ 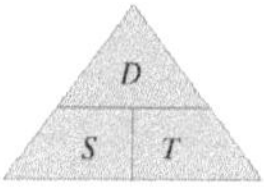
Inequalities on a graph	For $x > -3$, draw the graph of $x = -3$ with a dashed line; shade the region where $x > -3$ For $y \leqslant 4$, draw the graph of $y = 4$ with a solid line; shade the region where $y \leqslant 4$ For $y \geqslant mx + c$, draw the graph of $y = mx + c$ with a solid line; test points either side of the line to see which side to shade 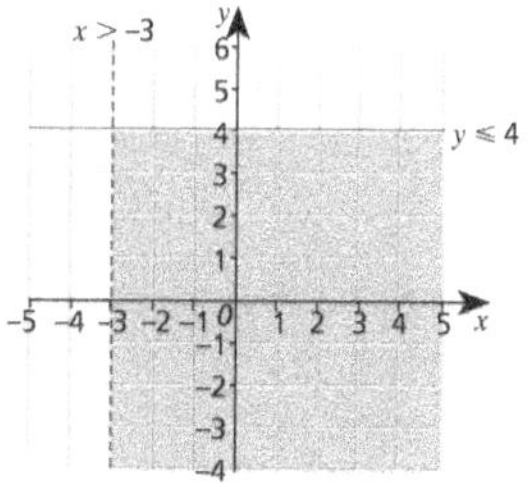
Inequalities on a number line	$x < 5$: 5 is not included 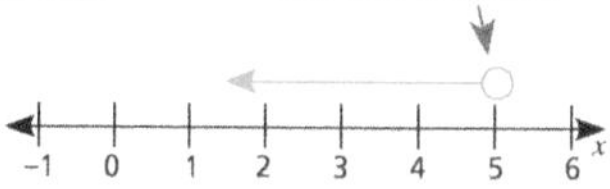 $2 < x \leqslant 4$: 4 is included 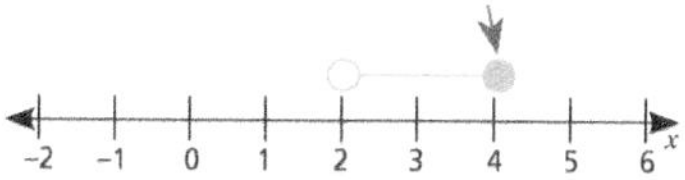
Quadratic graphs	Quadratic graph with a positive x^2 term 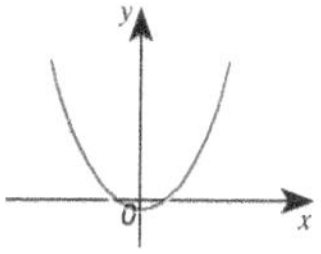 Quadratic graph with a negative x^2 term 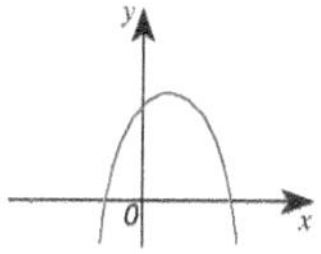 Quadratic graphs are symmetrical U-shaped curves
Solve	To solve an equation, find the value of the unknown letter When solving an equation, carry out the same operations to both sides at each step.
Solve an equation with brackets	Expand the brackets first
Solve an equation with fractions	To 'undo' the fraction, multiply both sides by the denominator When there are fractions on both sides, multiply both sides by the lowest common multiple (LCM) of their denominators
Solve an equation with x on both sides	Use inverse operations to get all the x terms on one side, and all the numbers on the other
Solve an inequality	Solve in the same way as an equation, by doing the same to both sides If you multiply or divide both sides by a negative number, reverse the inequality sign: $-y > 3$ $\times -1 \quad \times -1$ $y < -3$

Key facts and vocabulary

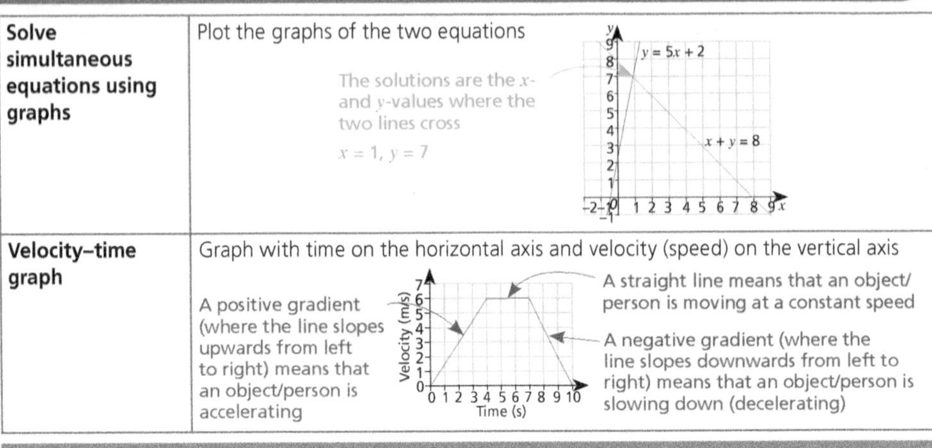

Solve simultaneous equations using graphs	Plot the graphs of the two equations The solutions are the x- and y-values where the two lines cross $x = 1$, $y = 7$
Velocity–time graph	Graph with time on the horizontal axis and velocity (speed) on the vertical axis A positive gradient (where the line slopes upwards from left to right) means that an object/person is accelerating A straight line means that an object/person is moving at a constant speed A negative gradient (where the line slopes downwards from left to right) means that an object/person is slowing down (decelerating)

Geometry

Congruent	Congruent shapes are exactly the same shape and size Two triangles are congruent if one or more of these four criteria are true: **Side, Side, Side (SSS)** · **Side, Angle, Side (SAS)** · **Angle, Side, Angle (ASA)** · **Right angle, Hypotenuse, Side (RHS)**
Cross section	A cross section is the 2D shape that is made when cutting through a 3D shape Cross section
Line of symmetry	Mirror line that divides a shape in half, so each half is a reflection of the other In a regular polygon, number of lines of symmetry = number of sides 1 2 3 4 5
Order of rotational symmetry	The number of times a shape 'lands on itself' when it is rotated a full turn In a regular polygon, the order of rotational symmetry = number of sides
Pythagoras' theorem	$c^2 = a^2 + b^2$ c is the length of the hypotenuse a and b are the lengths of the two shorter sides

Key facts and vocabulary

Scale factor	The number all side lengths are multiplied by in an enlargement A to B: enlargement scale factor 2 B to A: enlargement scale factor $\frac{1}{2}$
Similar	Similar shapes are enlargements of each other • Same angles • Same shape but different sizes • Each side length has been multiplied by the same scale factor
Surface area	The total area of all the faces of a 3D shape
Surface area of a cylinder	Surface area of a cylinder = $2\pi r^2 + 2\pi rh$
Trigonometry ratios	$\sin\theta = \frac{\text{opposite}}{\text{hypotenuse}} = \frac{\text{opp}}{\text{hyp}}$ $\cos\theta = \frac{\text{adjacent}}{\text{hypotenuse}} = \frac{\text{adj}}{\text{hyp}}$ $\tan\theta = \frac{\text{opposite}}{\text{adjacent}} = \frac{\text{opp}}{\text{adj}}$ You can remember these using S O H: $\sin = \frac{\text{opp}}{\text{hyp}}$ C A H: $\cos = \frac{\text{adj}}{\text{hyp}}$ T O A: $\tan = \frac{\text{opp}}{\text{adj}}$
Volume of a cylinder	Volume of a cylinder = area of circular cross section × height $= \pi r^2 h$
Volume of a prism	Volume of a prism = area of cross section × length

Probability

Biased	The outcomes are not equally likely; the opposite of 'fair'
Equally likely	All the outcomes have an equal probability of happening In this spinner, the events R, G, B and Y are all equally likely
Expected result	Expected result = P(event) × number of trials For rolling a dice 30 times, expected number of 3s = $\frac{1}{6} \times 30 = 5$

Key facts and vocabulary

Fair	Unbiased, e.g. a fair coin is equally likely to land on Heads or Tails
Independent events	The result of one event does not change the probability of the second event, e.g. for a dice, rolling a 6 does not change the probability of rolling a 6 next time
Mutually exclusive	Two events that cannot occur at the same time The probabilities of all the mutually exclusive events in a trial add up to 1 For this spinner P(Y) + P(G) + P(R) + P(B) = 1
Outcome	The result of a probability experiment or trial, e.g. for the experiment 'rolling a dice', possible outcomes are 1, 2, 3, 4, 5 or 6
Probability of an event	$\text{P(event)} = \frac{\text{number of ways the outcome can occur}}{\text{total number of possible outcomes}}$ For this spinner, $\text{P(R)} = \frac{1}{4}$
Probability of an event not happening	P(event not happening) = 1 – P(event happening) For this spinner P(not blue) = 1 – P(blue)
Probability scale	Impossible, Very unlikely, Unlikely, Even chance, Likely, Very likely, Certain 0 (0%) — $\frac{1}{2}$ (50%) — 1 (100%)
Relative frequency	$\text{Relative frequency} = \frac{\text{number of times event occurred}}{\text{total number of trials}}$ The greater the number of trials, the closer the relative frequency is to the theoretical probability
Sample space	Set of all possible outcomes, e.g. for the experiment 'rolling a dice', the sample space is 1, 2, 3, 4, 5, 6 A sample space for two combined events can be shown in a two-way table
Theoretical probability	Probability you calculate using this formula: $\text{P(event)} = \frac{\text{number of ways the outcome can occur}}{\text{total number of possible outcomes}}$
Venn diagram	A diagram showing the relationship between two or more things; this Venn diagram shows whether students had cereal, eggs or neither for breakfast Cereal: 12, Both: 3, Eggs: 8, Neither: 2

Answers

Page 5: Similarity

1. $9 \div 3 = 3$, so scale factor = 3
2. $12 \div 3 = 4$, so scale factor = 4
3. $2 \times x = 11$
   $x = \frac{11}{2}$
   So scale factor = $\frac{11}{2}$
4. $15 \times x = 3$
   $x = \frac{3}{15} = \frac{1}{5}$
   So scale factor = $\frac{1}{5}$
5. $5 \times x = 10$
   $x = 2$
   $2 \times 2 = 4$
   Unknown length = 4 cm
6. $6 \times x = 9$
   $x = \frac{9}{6} = \frac{3}{2}$
   $3 \times \frac{3}{2} = \frac{9}{2}$
   Unknown length = 4.5 cm
7. $4 \times x = 9$
   $x = \frac{9}{4}$
   $5 \div \frac{9}{4} = \frac{20}{9}$
   Unknown length = 2.2 cm

Page 7: Congruence

1. A and H B and G C and F D and E
2. No, they are not the same size.
3. No, the three side lengths are not equal.
4. Yes, the triangles are congruent. SSS is true.
5. Yes, the triangles are congruent. RHS is true.
6. No, the triangles are not congruent. The corresponding side is not the same.
7. Yes, the triangles are congruent. SAS is true.

Page 9: Symmetry

1. 1 line of symmetry

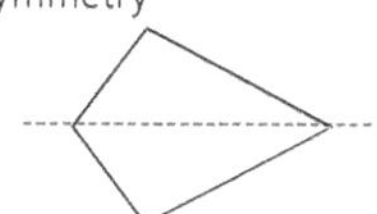

2. 5 lines of symmetry

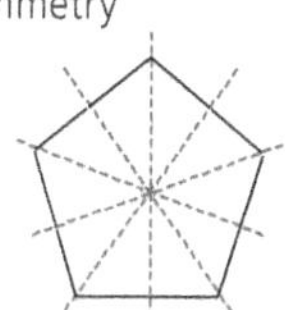

3. 1 line of symmetry

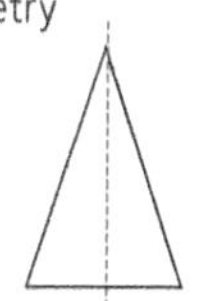

4. 4 lines of symmetry

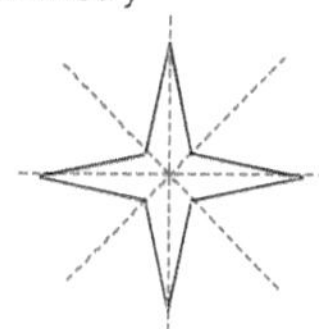

5. Order 2
6. Order 6
7. Order 1

Page 11: Pythagoras' theorem 1

1. **a)** 5 cm **b)** 11.2 cm (1 d.p.)
   c) 25 cm **d)** 7.5 cm (1 d.p.)
2. **a)** 8.5 cm (1 d.p.) **b)** 11.3 mm (1 d.p.)
   c) 4.5 m (1 d.p.) **d)** 3.4 km (1 d.p.)
3. No $10^2 \neq 5^2 + 7^2$
4. Yes $17^2 = 8^2 + 15^2$

Page 13: Pythagoras' theorem 2

1. **a)** 15 cm **b)** 15.2 cm (1 d.p.)
   c) 14.5 cm (1 d.p.) **d)** 6.4 cm (1 d.p.)
2. 77 mm
3. **a)** 6.4 cm (1 d.p.) **b)** 18.4 m (1 d.p.)
4. 14.4 cm (1 d.p.)
5. **a)** No $15^2 \neq 10^2 + 7^2$
   b) Yes $29^2 = 20^2 + 21^2$

Page 15: Solving problems using Pythagoras' theorem

1. 3.03 m
2. **a)** 27 inches **b)** 55 inches
3. 2.75 m
4. 13 cm
5. **a)** 4 cm **b)** 20 cm^2
   c) AB = 8.1 cm (1 d.p.)
   Perimeter = 23.1 cm (1 d.p.)

Page 17: Introducing probability

1. Options should be matched as follows:

Event	to	Spinning red or blue
Outcome	to	Spinning yellow
Sample space	to	Red, yellow and blue
Trial	to	Spinning the spinner
Unbiased	to	Each section is equal

2. a) Even chance – there are 26 red cards and 26 black cards
   b) Unlikely – there is one 1 out of 10 total numbers
   c) Certain
   d) Impossible
3. She is incorrect. There is a $\frac{1}{6}$ chance of rolling a 6.
4. a) E is more likely; it is closer to 1 than B
   b) B, E and A are all less likely as they are closer to 0
   c) A Evens; B Unlikely; C Certain; D Likely; E Unlikely

Page 19: Probability of single events

1. a) S = {1, 2, 3, 4, 5, 6}
   b) S = {M, A, T, H, E, M, A, T, I, C, S}
   c) S = {R, R, R, B, B, Y, G, G, G, G}
2. a) $P(\leq 4) = \frac{4}{6} = \frac{2}{3}$
   b) $P(M) = \frac{2}{11}$
   c) $P(G) = \frac{4}{10} = \frac{2}{5}$
3. 100% – 52% = 48%
4. a) $P(6) = \frac{1}{6}$
   b) $\frac{1}{6} \times 24 = 4$
   c) Relative frequency = $\frac{\text{number of times 6 is rolled}}{\text{number of times dice is rolled}}$
   $= \frac{2}{24}\ (= \frac{1}{12})$

Page 21: Calculating probabilities of combined events

1. a)

Spinner 1 \ Spinner 2	1	2	3	4
1	2	3	4	5
2	3	4	5	6
3	4	5	6	7
4	5	6	7	8

   b) H1, H2, H3, H4, T1, T2, T3, T4

Coin \ Spinner	1	2	3	4
H	H1	H2	H3	H4
T	T1	T2	T3	T4

2. a) There are 3 ways to get a sum of 4 out of $4 \times 4 = 16$ possible outcomes
   $P(\text{sum of } 4) = \frac{3}{16}$
   b) There are 13 ways to get a sum of greater than 3 out of $4 \times 4 = 16$ possible outcomes
   $P(\text{sum greater than } 3) = \frac{13}{16}$
3. a) There are 2 ways to get a Tails and an even number out of $2 \times 4 = 8$ possible outcomes
   $P(\text{T \& even}) = \frac{2}{8} = \frac{1}{4}$
   b) The probabilities of all the mutually exclusive events sum to 1.
   P(NOT (T & even)) = 1 – P(T & even)
   $= 1 - \frac{1}{4} = \frac{3}{4}$

Page 23: Venn diagrams

1. Options should be matched as follows:
   Set A and Set B to the left-hand Venn diagram
   Neither Set A nor Set B to the middle Venn diagram
   Set A to the right-hand Venn diagram
2. a) 50 workers were surveyed. Add up all the numbers in the diagram.
   12 + 18 + 7 + 13 = 50
   b) 25 workers took the train. Add up the number in the circle for train and in the intersection.
   18 + 7 = 25
   c) 7 workers took the bus and a train. This is the number in the intersection of train and bus.
   d) 12 workers took neither a bus nor a train. This is the number outside the circles.
3.

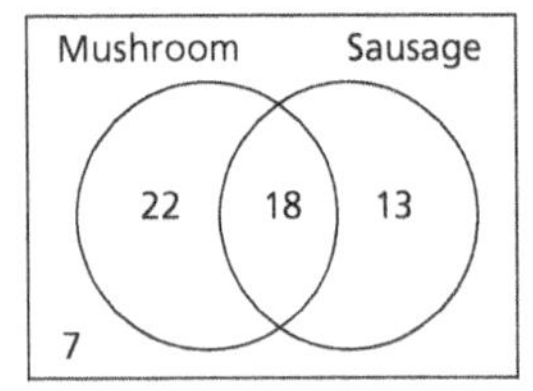

Page 25: Geometric and special number sequences

1. a) Not geometric (it is arithmetic)
   b) Geometric (it is dividing by 5 each time)
   c) Not geometric (it doesn't have a common ratio)
   d) Geometric (it is multiplying by 10 each time)
2. 2, 6, 18, 54, 162
3. a) First, find the common ratio $32 \div 8 = 4$, then the missing term is $32 \times 4 = 128$

b) First, find the common ratio $4 \div 16 = \frac{1}{4}$, then the missing term is $256 \times \frac{1}{4}$

$= 256 \div 4 = 64$

4. The first triangle has 15 dots, the next has 21 dots and the next has 28 dots, so the sequence is increasing by + 6 then + 7. The next triangle will have 28 + 8 = 36 dots

5. The sequence is increasing in the pattern + 3, + 5, + 7, + 9 so the next term will be + 11.

27 + 11 = 38

Page 27: Expanding two or more binomials

1. $(12 \times 10) + (12 \times 20) = 12 \times 30$

$120 + 240 = 360$

2. a)

	x	-4
$4x$	$x \times 4x = 4x^2$	$-4 \times 4x = -16x$
$+3$	$x \times 3 = 3x$	$-4 \times 3 = -12$

$4x^2 - 16x + 3x - 12$

$= 4x^2 - 13x - 12$

b)

	$2n$	-1
$3n$	$2n \times 3n = 6n^2$	$-1 \times 3n = -3n$
-2	$2n \times -2 = -4n$	$-1 \times -2 = 2$

$6n^2 - 3n - 4n + 2$

$= 6n^2 - 7n + 2$

3. a) $(4y + 2)(y - 5) = 4y^2 - 20y + 2y - 10$

$= 4y^2 - 18y - 10$

b) $(2k + 3)(3k + 5) = 6k^2 + 10k + 9k + 15$

$= 6k^2 + 19k + 15$

4. $(x - 4)(4x + 3)(x - 2) = (4x^2 - 13x - 12)(x - 2)$

$= 4x^3 - 8x^2 - 13x^2 + 26x - 12x + 24$

$= 4x^3 - 21x^2 + 14x + 24$

Page 29: Solving multi-step equations 1

1. a) $7x + 9 = 51$

$7x = 42$

$x = 6$

b) $6x - 9 = 21$

$6x = 30$

$x = 5$

2. a) $2x + 8 = 4x - 4$

$8 = 2x - 4$

$12 = 2x$

$x = 6$

b) $6x - 1 = 15 - 2x$

$8x - 1 = 15$

$8x = 16$

$x = 2$

3. a) $3(x - 5) + 2(9 - x) = 10$

$3x - 15 + 18 - 2x = 10$

$x + 3 = 10$

$x = 7$

b) $6x + 7(2x - 3) = 19$

$6x + 14x - 21 = 19$

$20x - 21 = 19$

$20x = 40$

$x = 2$

Page 31: Solving multi-step equations 2

1. $3x = -6$ should be $-3x = -6$

They could first manipulate the equation to get the coefficient of x to be positive.

$4 - 3x = -2$

$4 = -2 + 3x$

$6 = 3x$

$x = 2$

2. a) $\frac{3k - 2}{4} = 4$

$\frac{3k - 2}{4} \times 4 = 4 \times 4$

$3k - 2 = 16$

$3k = 18$

$k = 6$

b) $2 + \frac{x + 1}{3} = 5$

$(2 \times 3) + (\frac{x + 1}{3} \times 3) = 5 \times 3$

$6 + x + 1 = 15$

$x + 7 = 15$

$x = 8$

3. $\frac{3 - 2m}{2} = \frac{m + 2}{6}$

$\frac{3 - 2m}{2} \times 6 = \frac{m + 2}{6} \times 6$ (LCM of 2 and 6 is 6)

$3(3 - 2m) = m + 2$

$9 - 6m = m + 2$

$9 = 7m + 2$

$7 = 7m$

$m = 1$

Page 33: Solving inequalities

1. a) Any five values less than 7, not including 7, e.g. 6, 5, 4, 3, 2

b) Any five values greater than or equal to 25, e.g. 25, 26, 27, 28, 29

c) Any five values between 5 and 10, including 5 and 10, e.g. 5, 6, 7, 8, 9

2. a)

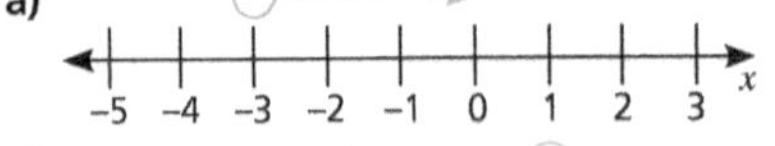

b)

−1 0 1 2 3 4 5 6 7 8 x

c)

−7 −6 −5 −4 −3 −2 −1 0 1 2 x

3. a) $4k + 2 < 10$

$4k < 8$

$k < 2$

b) $7 - 8m \geqslant -9$

$7 \geqslant -9 + 8m$

$16 \geqslant 8m$

$2 \geqslant m$

$m \leqslant 2$

c) $4 < 10 - 3j$

$4 + 3j < 10$

$3j < 6$

$j < 2$

Page 35: Using and writing formulae

1. Example answers:

a) In words: The perimeter of a regular polygon is the side length multiplied by the number of sides.

Algebraically: $P = sl$, where P = perimeter, s = number of sides and l = length of the side

b) In words: The cost of the party is £100 plus the number of guests multiplied by £12.

Algebraically: $C = 100 + 12g$, where C = the cost of the party in £ and g = the number of guests

2. a) $P = sl$, then substitute $s = 10$ and $l = 1.3$

$P = 10 \times 1.3 = 13$ cm

b) $C = 100 + 12g$, then substitute $g = 25$

$C = 100 + (12 \times 25) = £400$

3. Using the formula triangle for $S = DT$, then $T = D \div S$

Substituting $D = 630$ km and $S = 280$, then $T = 630 \div 280 = 2.25$ hours

Page 37: Rearranging formulae

1. a) P b) A c) C

2. a) $P = sl$

$P \div s = l$

$l = \frac{P}{s}$

b) Substituting $P = 108$ and $s = 12$ into $l = \frac{P}{s}$

$l = \frac{108}{12} = 9$

3. a) $A = \pi r^2$

$\div \pi \quad \div \pi$

$\frac{A}{\pi} = r^2$

$\sqrt{\frac{A}{\pi}} = \sqrt{r^2}$

$r = \sqrt{\frac{A}{\pi}}$

b) Substituting $A = 225\pi$, then $r = \sqrt{\frac{225\pi}{\pi}}$

$= \sqrt{225} = 15$ cm

4. a) $C = 30h + 45$

$C - 45 = 30h$

$(C - 45) \div 30 = h$

$h = \frac{C - 45}{30}$

b) Substituting $C = 135$, then

$h = \frac{135 - 45}{30} = \frac{90}{30} = 3$ hours

Page 39: Trigonometry 1

1. a) $x = 10.4$ cm (1 d.p.) b) $y = 8$ cm
   c) $z = 7.3$ cm (1 d.p.) d) $n = 5.3$ cm (1 d.p.)
2. a) $a = 38.6$ cm (1 d.p.) b) $b = 4.7$ cm (1 d.p.)
3. a) 45° and 45°
   b) 7.1 cm and 7.1 cm (1 d.p.)
4. a) 30° b) 54.9° (1 d.p.)
   c) 36.9° (1 d.p.) d) 22.6° (1 d.p.)

Page 41: Trigonometry 2

1. a) $x = 7.7$ cm (1 d.p.) b) $y = 9.2$ cm (1 d.p.)
2. a) $r = 13.7$ cm (1 d.p.) b) $t = 4.2$ cm (1 d.p.)
3. a) $m = 9.7$ cm (1 d.p.) b) $n = 9.9$ cm (1 d.p.)
4. a) 60.9° (1 d.p.) b) 45.6° (1 d.p.)
   c) 40.6° (1 d.p.) d) 63.4° (1 d.p.)

Page 43: Solving problems involving trigonometry

1. 5.77 m (to the nearest cm)
2. 2.72 m (to the nearest cm)
3. 5.6 cm (1 d.p.)
4. 59.7° (1 d.p.)
5. a) 2.2 cm (1 d.p.) b) 5.2 cm (1 d.p.)

Page 45: Solving problems involving right-angled triangles

1. 0.92 m or 92 cm (to the nearest cm)
2. a) 29.73 m (2 d.p.) b) 82.3° (1 d.p.)

3. **a)** $20.6°$ (1 d.p.) **b)** 8.5 cm (1 d.p.)
4. 4.2 cm (1 d.p.)
5. **a)** $22.6°$ (1 d.p.) **b)** $13.0°$ (1 d.p.)
   c) 13.3 cm (1 d.p.) **d)** 12 cm

Page 47: Properties of 3D shapes 1

1. A face is a flat or curved surface on a 3D shape.
2. An edge is the line where two faces join.
3. A vertex is a corner where edges meet.
4. Cube or cuboid
5. Cone
6. Square-based pyramid
7. Cylinder
8. Triangular prism

Page 49: Properties of 3D shapes 2

1. **a)** Area of one face: $5 \times 5 = 25\,\text{cm}^2$
   Area of six faces: $25 \times 6 = 150\,\text{cm}^2$
   b) $5 \times 5 \times 5 = 125\,\text{cm}^3$
2. **a)** $6 + 6 + 18 + 24 + 30 = 84\,\text{cm}^2$
   b) $\frac{4 \times 3}{2} = 6\,\text{cm}^2$
   $6 \times 6 = 36\,\text{cm}^3$
3. **a)** Area of one circle: $5^2 \times \pi = 25\pi\text{cm}^2$
   Area of two circles: $25\pi \times 2 = 50\pi\text{cm}^2$
   Area of rectangle:
   Circumference of circle $= \pi \times 10 = 10\pi$
   $10\pi \times 12 = 120\pi\text{cm}^2$
   Total surface area $= 50\pi\text{cm}^2 + 120\pi\text{cm}^2 = 170\pi\text{cm}^2$
   b) Volume $= \pi r^2 \times h$
   $= \pi(5)^2 \times 12$
   $= 300\pi\text{cm}^3$
4. Volume $= \frac{1}{2}\pi r^2 h + lwh$
   $= \frac{1}{2}\pi(30)^2(90) + (15)(60)(90)$
   $= 208\,234.5025$
   $= 210\,000\,\text{cm}^3$

Page 51: Solving problems in 3D

1. **a)** Surface area of one face is $54 \div 6 = 9\,\text{cm}^2$
   So side length of the cube is $\sqrt{9} = 3\,\text{cm}$
   Total surface area of cuboid is:
   $((9 \times 3) \times 2) + ((6 \times 3) \times 2) + ((6 \times 9) \times 2)$
   $54 + 36 + 108 = 198\,\text{cm}^2$
   b) $9 \times 3 \times 6 = 162\,\text{cm}^3$
2. **a)** Surface area of one face is $96 \div 6 = 16\,\text{cm}^2$
   So side length of the cube is $\sqrt{16} = 4\,\text{cm}$
   Total surface area of cuboid is:
   $((24 \times 4) \times 4) + (4 \times 4) \times 2))$
   $384 + 32 = 416\,\text{cm}^2$
   b) $24 \times 4 \times 4 = 384\,\text{cm}^3$
3. **a)** Volume of the metal block is
   $25 \times 30 \times 25 = 18\,750\,\text{cm}^3$
   Volume of the cylinder is $(4)^2 \times \pi \times 30$
   $= 480\pi\text{cm}^3$
   Volume of the block after the hole has been drilled is: $18\,750\,\text{cm}^3 - 480\pi\text{cm}^3$
   $= 17\,242\,\text{cm}^3$
   b) $\frac{480\pi}{18750} \times 100 = 8.042...\%$ or 8%

Page 53: Introducing standard form

1. **a)** 100 000 000 **b)** 0.0001
   c) 1 000 000 **d)** 0.000 000 01
2. 1.234×10^{-10} 1.2×10^{34} 1×10^{23} 4×10^{0}
3. **a)** 5.326×10^{3} **b)** 7.2835×10^{7}
   c) 7.26×10^{-3} **d)** 2.09×10^{-1}
4. **a)** 112 000 000 **b)** 0.000 618
   c) 4 190 000 **d)** 0.00205

Page 55: Comparing and ordering numbers in standard form

1. **a)** $1.35 \times 10^{8} < 1.38 \times 10^{10}$
   b) $6.95 \times 10^{5} > 6.9 \times 10^{5}$
   c) $9 \times 10^{2} < 9 \times 10^{3}$
   d) $6.239 \times 10^{6} > 1.53 \times 10^{5}$
2. **a)** $4.09 \times 10^{8} > 4.09 \times 10^{-10}$
   b) $9.64 \times 10^{-5} > 9.6 \times 10^{-5}$
   c) $1.8 \times 10^{-5} < 1.8 \times 10^{-2}$
   d) $4.21 \times 10^{-6} < 2.592 \times 10^{2}$
3. $8.06 \times 10^{6} > 7.8 \times 10^{6} > 2.64 \times 10^{0} > 3.03 \times 10^{-7} > 2.6 \times 10^{-9}$

Page 57: Distance–time graphs

1. **a)** 9 km
   b) $1\frac{1}{2}$ hours (90 minutes)
   c) 9 km/h
   d) 4 hours
   e) On the way to her friend's – this line is steeper.
   Or compare the times – on the way there she cycles 9 km in 1 hour. On the way back, she cycles 9 km in $1\frac{1}{2}$ hours, which is slower.
   f) 18 km

2. a)

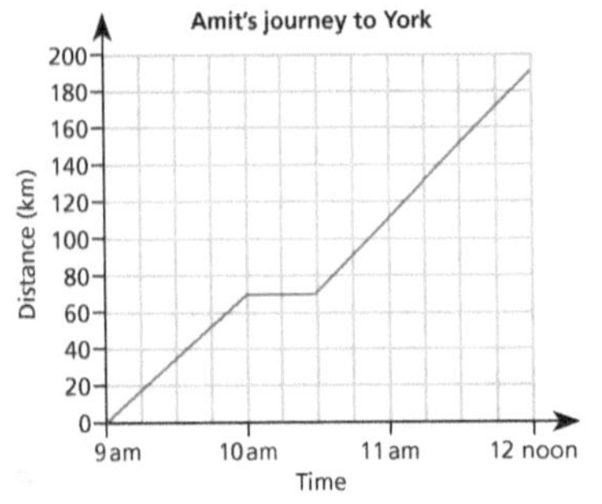

b) After his break **c)** 190 km

3. a) Bus A **b)** $1\frac{1}{2}$ hours (90 mins)

c) 70 km **d)** Bus B

Page 59: Interpreting other real-life graphs

1. From 0 to 10 seconds, the car is accelerating.

The car then travels at a constant speed between 10 and 20 seconds but it begins to accelerate again between 20 and 30 seconds.

The car then slows down for the final 20 seconds before it stops at 50 seconds.

2. a) £40 **b)** $30

Page 61: Plotting real-life graphs

1.

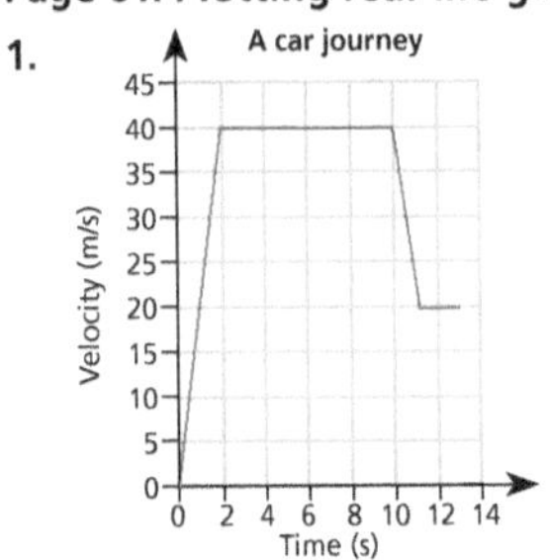

2. a)

Distance (miles)	0	1	2	4	6	8	10
Total cost (£)	3.00	3.50	4.00	5.00	6.00	7.00	8.00

b)

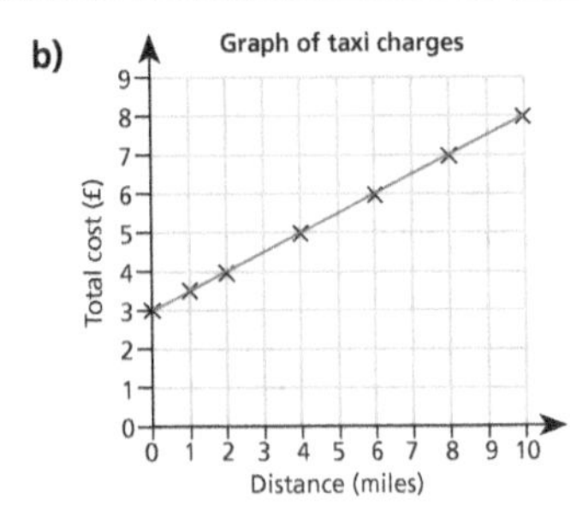

3. a)

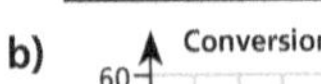

Pounds (£)	1.00	10.00	20.00	25.00	50.00
Euros (€)	1.20	12.00	24.00	30.00	60.00

b)

Conversion graph for £ and €

Euros (€)

Pounds (£)

4. a)

Inches	2	4	12	20	25
Centimetres	5	10	30	50	62.5

b)

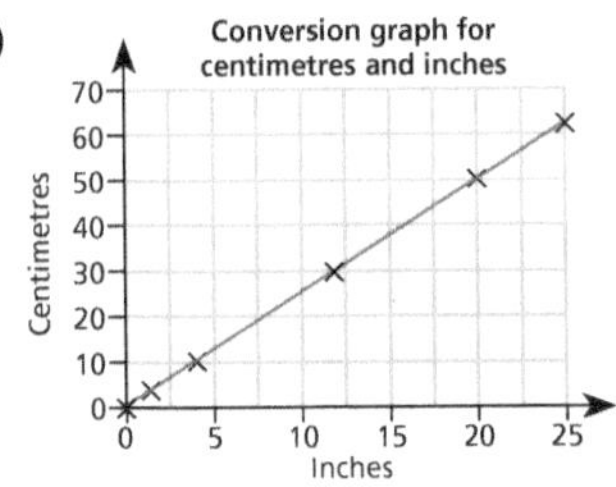

Page 63: Rearranging linear equations to $y = mx + c$

1. a) $2y - 10x = 6$

$2y = 10x + 6$

$y = 5x + 3$

b) Gradient = 5, y-intercept = 3

2. a) $2y + 8 = 6x$

$2y = 6x - 8$

$y = 3x - 4$

b) Gradient = 3, y-intercept = −4

3. a) $2y + 2 = 4x$

$2y = 4x - 2$

$y = 2x - 1$

b) Gradient = 2, y-intercept = −1

4. a) $4y + 3 = 4x - 5$

$4y = 4x - 8$

$y = x - 2$

b) Gradient = 1, y-intercept = −2

5. a) $y - 2 = -4x + 7$

$y = -4x + 9$

b) Gradient = −4, y-intercept = 9

6. $3x + y = 10$

   $y = -3x + 10$

7. $4x + 2y = 6$

   $2y = -4x + 6$

   $y = -2x + 3$

Page 65: Using graphs to solve problems

1. **a) i)** £2.90 **ii)** £4

   b)

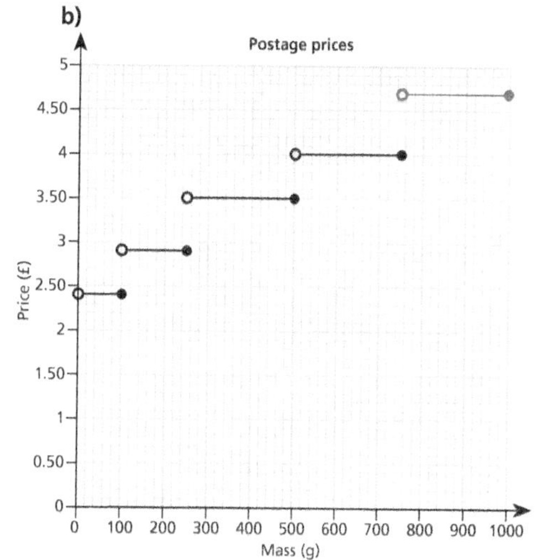

2. **a)** A value between 41 and 43 cells

   b) A value between 4.6 and 4.8 hours

Page 67: Using graphs to solve simultaneous linear equations

1. **a)** $x = 0$, $y = 5$

   b) $x = 1$, $y = 4$

   c) $x = 3$, $y = 8$

2.

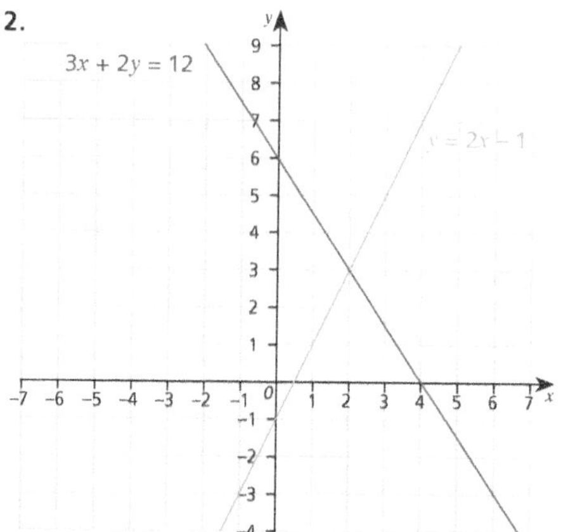

$x = 2$, $y = 3$

Page 69: Graphing inequalities

1. **a) and b)**

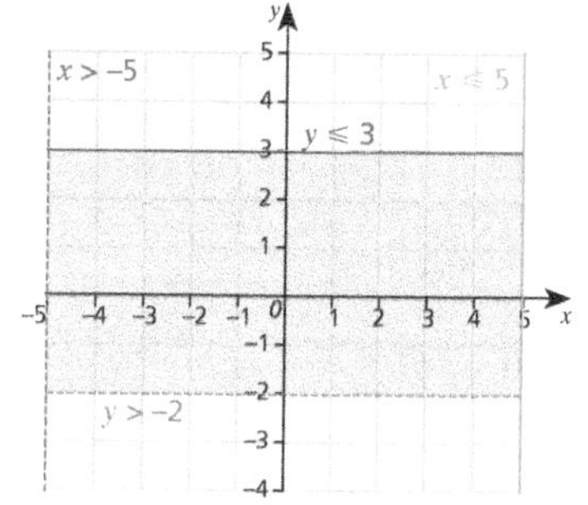

2.

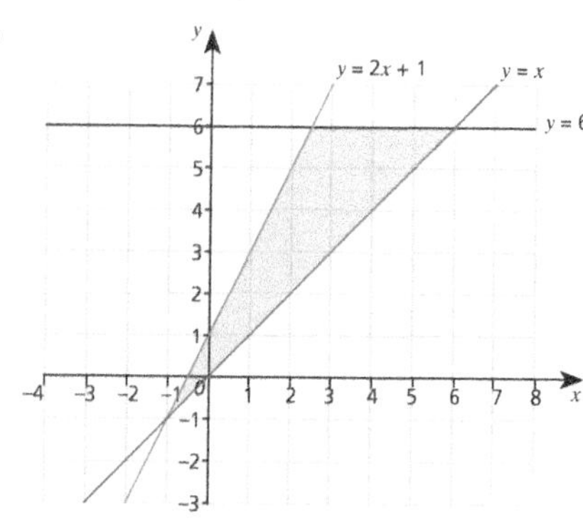

3.

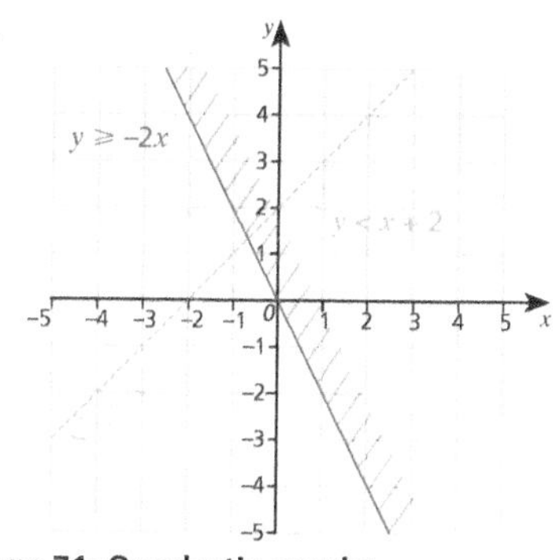

Page 71: Quadratic graphs

1. **a)**

x	–3	–2	–1	0	1	2	3
x^2	9	4	1	0	1	4	9
$y = -x^2$	–9	–4	–1	0	–1	–4	–9

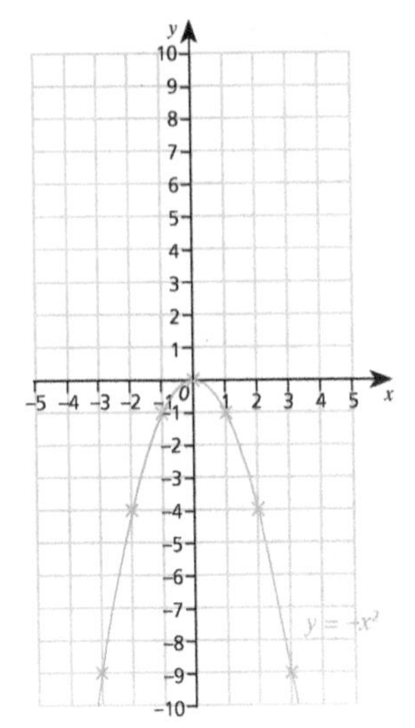

b)

x	−3	−2	−1	0	1	2	3
x^2	9	4	1	0	1	4	9
$y = x^2 + 1$	10	5	2	1	2	5	10

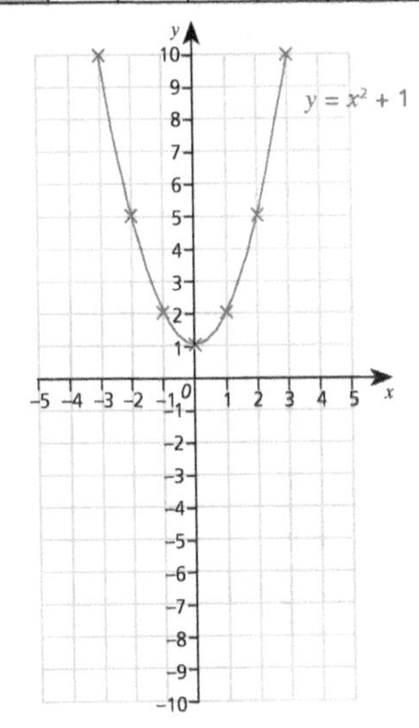

2.

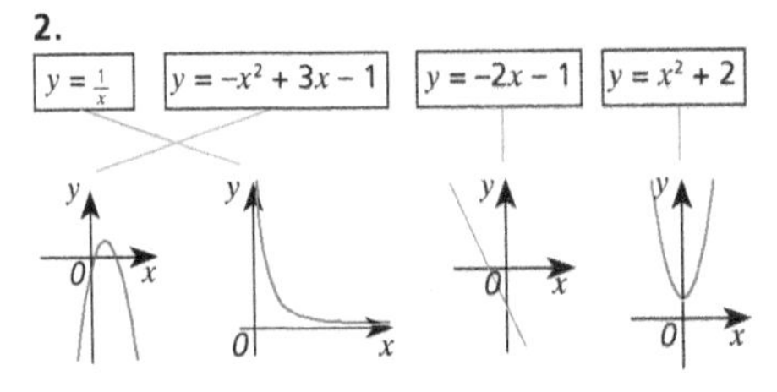

Pages 72–75: Mixed questions

1. **a)** 8 **b)** 2
2. **a)** $\frac{1}{2}$ **b)** $\frac{3}{13}$ **c)** $\frac{1}{52}$
3. **a)** $3x^2 + 5x - 2$ **b)** $6x^3 - 5x^2 - 29x + 10$
4. **a)** $a = -\frac{11}{4}$ **b)** $b = 13$
5. Yes, the triangles are congruent.
   First triangle: 80° + 40° = 120°
   Unmarked angle is 180° − 120° = 60°
   Second triangle: 80° + 60° = 140°
   Unmarked angle is 180° − 140° = 40°
   So corresponding angles are equal and the corresponding side is equal. ASA is true.
6. **a)** $x < 6$ **b)** $x = 1, 2, 3, 4, 5$
7. 10 cm
8. **a)** $h = \frac{V}{\pi r^2}$ **b)** 3.82 cm
9. 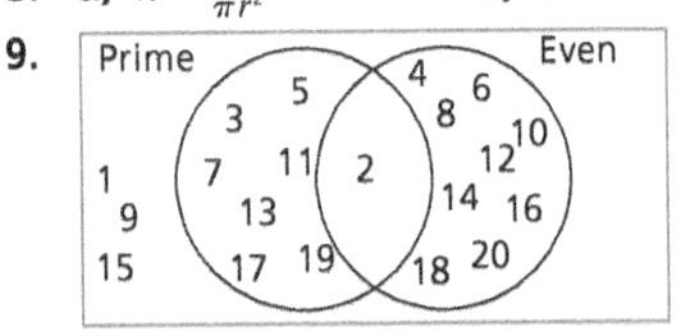

10. **a)** 10.1 m **b)** 36.5°
11. **a)** 3.844×10^5 km **b)** 150 000 000 km
12. **a)** $y - 3x = 10$ **b)** $4x - y - 3 = 0$
13. **a)** 25 000 cm³ **b)** 200
14. **a)** 78 cm **b)** 2.9 m
15. $x = 2$, $y = 3$
16. **a)** 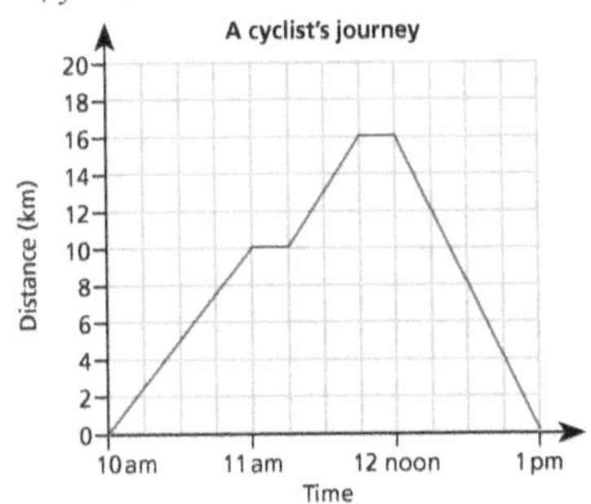

    b) 16 km/h
17. **a)** 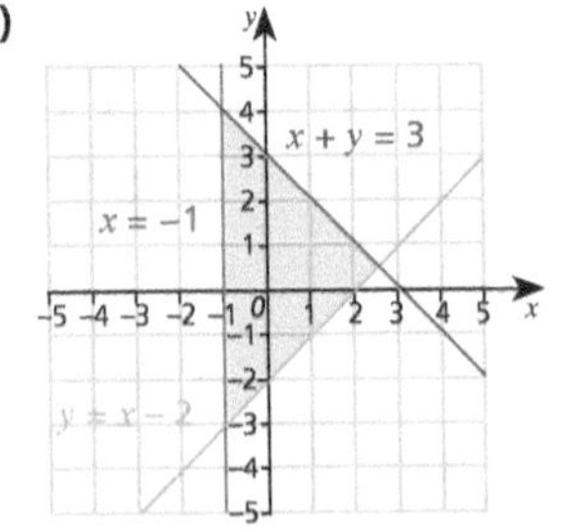

    b) (1, 1)